崧燁文化

曹永忠、許智誠、蔡英德　著

Arduino程式教學
(語音模組篇)

Arduino Programming (Voice Modules)

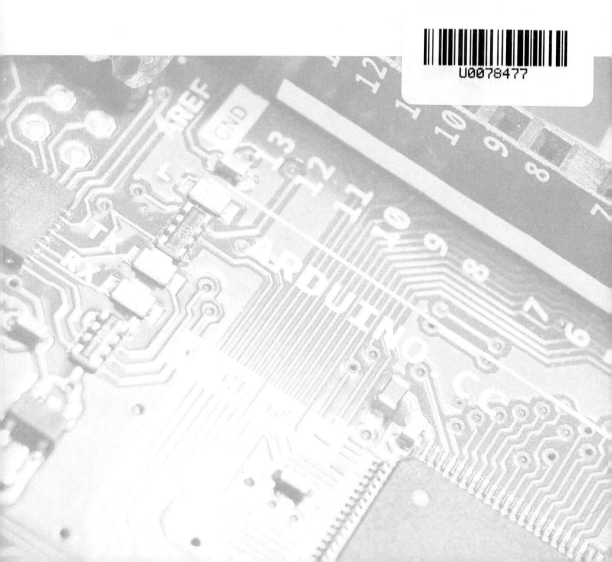

自序

Arduino 系列的書出版至今，已經兩年半多，出書量也破七十本大關，當初出版電子書是希望能夠在教育界開一門 Maker 自造者相關的課程，沒想到一寫就已過三年，繁簡體加起來的出版數也已破七十本的量，這些書都是我學習當一個 Maker 累積下來的成果。

這本書可以說是我的書另一個里程碑，很久以前，這個系列開始以駭客的觀點為主，希望 Maker 可以擁有駭客的觀點、技術、能力，駭入每一個產品設計思維，並且成功的重製、開發、超越原有的產品設計，這才是一位對社會有貢獻的『駭客』。

如許多學習程式設計的學子，為了最新的科技潮流，使用著最新的科技工具與軟體元件，當他們面對許多原有的軟體元件沒有支持的需求或軟體架構下沒有直接直持的開發工具，此時就產生了莫大的開發瓶頸，這些都是為了追求最新的科技技術而忘卻了學習原有基礎科技訓練所致。

筆著鑒於這樣的困境，思考著『如何駭入眾人現有知識寶庫轉換為我的知識』的思維，如果我們可以駭入產品結構與設計思維，那麼了解產品的機構運作原理與方法就不是一件難事了。更進一步我們可以將原有產品改造、升級、創新，並可以將學習到的技術運用其他技術或新技術領域，透過這樣學習思維與方法，可以更快速的掌握研發與製造的核心技術，相信這樣的學習方式，會比起在已建構好的開發模組或學習套件中學習某個新技術或原理，來的更踏實的多。

目前許多學子在學習程式設計之時，恐怕最不能了解的問題是，我為何要寫九九乘法表、為何要寫遞迴程式，為何要寫成函式型式…等等疑問，只因為在學校的學子，學習程式是為了可以了解『撰寫程式』的邏輯，並訓練且建立如何運用程式邏輯的能力，解譯現實中面對的問題。然而現實中的問題往往太過於複雜，授課的老師無法有多餘的時間與資源去解釋現實中複雜問題，期望能將現實中複雜問題淬鍊成邏輯上的思路，加以訓練學生其解題思路，但是眾多學子宥於現實問題的困惑，無法單純用純粹的解題思路來進行學習與訓練，反而以現實中的複雜來反駁老師教學太過學理，沒有實務上的應用為由，拒絕深入學習，這樣的情形，反而自己造成了學習上的障礙。

本系列的書籍，針對目前學習上的盲點，希望讀者當一位產品駭客，將現有產品的產品透過逆向工程的手法，進而了解核心控制系統之軟硬體，再透過簡單易學的 Arduino 單晶片與 C 語言，重新開發出原有產品，進而改進、加強、創新其原有產品固有思維與架構。如此一來，因為學子們進行『重新開發產品』過程之中，可以很有把握的了解自己正在進行什麼，對於學習過程之中，透過實務需求導引著開發過程，可以讓學子們讓實務產出與邏輯化思考產生關連，如此可以一掃過去陰霾，更踏實的進行學習。

這三年多以來的經驗分享，逐漸在這群學子身上看到發芽，開始成長，覺得 Maker 的教育方式，極有可能在未來成為教育的主流，相信我每日、每月、每年不斷的努力之下，未來 Maker 的教育、推廣、普及、成熟將指日可待。

最後，請大家可以加入 Maker 的 Open Knowledge 的行列。

曹永忠 於貓咪樂園

自序

記得自己在大學資訊工程系修習電子電路實驗的時候,自己對於設計與製作電路板是一點興趣也沒有,然後又沒有天分,所以那是苦不堪言的一堂課,還好當年有我同組的好同學,努力的照顧我,命令我做這做那,我不會的他就自己做,如此讓我解決了資訊工程學系課程中,我最不擅長的課。

當時資訊工程學系對於設計電子電路課程,大多數都是專攻軟體的學生去修習時,系上的用意應該是要大家軟硬兼修,尤其是在台灣這個大部分是硬體為主的產業環境,但是對於一個軟體設計,但是缺乏硬體專業訓練,或是對於眾多機械機構與機電整合原理不太有概念的人,在理解現代的許多機電整合設計時,學習上都會有很多的困擾與障礙,因為專精於軟體設計的人,不一定能很容易就懂機電控制設計與機電整合。懂得機電控制的人,也不一定知道軟體該如何運作,不同的機電控制或是軟體開發常常都會有不同的解決方法。

除非您很有各方面的天賦,或是在學校巧遇名師教導,否則通常不太容易能在機電控制與機電整合這方面自我學習,進而成為專業人員。

而自從有了 Arduino 這個平台後,上述的困擾就大部分迎刃而解了,因為Arduino 這個平台讓你可以以不變應萬變,用一致性的平台,來做很多機電控制、機電整合學習,進而將軟體開發整合到機構設計之中,在這個機械、電子、電機、資訊、工程等整合領域,不失為一個很大的福音,尤其在創意掛帥的年代,能夠自己創新想法,從 Original Idea 到產品開發與整合能夠自己獨立完整設計出來,自己就能夠更容易完全了解與掌握核心技術與產業技術,整個開發過程必定可以提供思維上與實務上更多的收穫。

Arduino 平台引進台灣自今,雖然越來越多的書籍出版,但是從設計、開發、製作出一個完整產品並解析產品設計思維,這樣產品開發的書籍仍然鮮見,尤其是能夠從頭到尾,利用範例與理論解釋並重,完完整整的解說如何用 Arduino 設計出一個完整產品,介紹開發過程中,機電控制與軟體整合相關技術與範例,如此的書籍更是付之闕如。永忠、英德兄與敝人計畫撰寫 Maker 系列，就是基於這樣對市場需要的觀察，開發出這樣的書籍。

　　作者出版了許多的 Arduino 系列的書籍，深深覺的，基礎乃是最根本的實力，所以回到最基礎的地方，希望透過最基本的程式設計教學，來提供眾多的 Makers 在入門 Arduino 時，如何開始，如何攬寫自己的程式，進而介紹不同的週邊模組，主要的目的是希望學子可以學到如何使用這些週邊模組來設計程式，期望在未來產品開發時，可以更得心應手的使用這些週邊模組與感測器，更快將自己的想法實現，希望讀者可以了解與學習到作者寫書的初衷。

　　　　　　　　　　　　　許智誠　　於中壢雙連坡中央大學 管理學院

自序

隨著資通技術(ICT)的進步與普及，取得資料不僅方便快速，傳播資訊的管道也多樣化與便利。然而，在網路搜尋到的資料卻越來越巨量，如何將在眾多的資料之中篩選出正確的資訊，進而萃取出您要的知識？如何獲得同時具廣度與深度的知識？如何一次就獲得最正確的知識？相信這些都是大家共同思考的問題。

為了解決這些困惱大家的問題，永忠、智誠兄與敝人計畫製作一系列「Maker系列」書籍來傳遞兼具廣度與深度的軟體開發知識，希望讀者能利用這些書籍迅速掌握正確知識。首先規劃「以一個 Maker 的觀點，找尋所有可用資源並整合相關技術，透過創意與逆向工程的技法進行設計與開發」的系列書籍，運用現有的產品或零件，透過駭入產品的逆向工程的手法，拆解後並重製其控制核心，並使用 Arduino 相關技術進行產品設計與開發等過程，讓電子、機械、電機、控制、軟體、工程進行跨領域的整合。

近年來 Arduino 異軍突起，在許多大學，甚至高中職、國中，甚至許多出社會的工程達人，都以 Arduino 為單晶片控制裝置，整合許多感測器、馬達、動力機構、手機、平板...等，開發出許多具創意的互動產品與數位藝術。由於 Arduino 的簡單、易用、價格合理、資源眾多，許多大專院校及社團都推出相關課程與研習機會來學習與推廣。

以往介紹 ICT 技術的書籍大部份以理論開始、為了深化開發與專業技術，往往忘記這些產品產品開發背後所需要的背景、動機、需求、環境因素等，讓讀者在學習之間，不容易了解當初開發這些產品的原始創意與想法，基於這樣的原因，一般人學起來特別感到吃力與迷惘。

本書為了讀者能夠深入了解產品開發的背景，本系列整合 Maker 自造者的觀念與創意發想，深入產品技術核心，進而開發產品，只要讀者跟著本書一步一步研習與實作，在完成之際，回頭思考，就很容易了解開發產品的整體思維。透過這樣的思路，讀者就可以輕易地轉移學習經驗至其他相關的產品實作上。

所以本書是能夠自修的書，讀完後不僅能依據書本的實作說明準備材料來製作，盡情享受 DIY(Do It Yourself)的樂趣，還能了解其原理並推展至其他應用。有興趣的讀者可再利用書後的參考文獻繼續研讀相關資料。

　　本書的發行有新的創舉，就是以電子書型式發行，在國家圖書館(http://www.ncl.edu.tw/)、國立公共資訊圖書館 National Library of Public Information(http://www.nlpi.edu.tw/)、台灣雲端圖庫(http://www.ebookservice.tw/)等都可以免費借閱與閱讀，如要購買的讀者也可以到許多電子書網路商城、Google Books 與 Google Play 都可以購買之後下載與閱讀。希望讀者能珍惜機會閱讀及學習，繼續將知識與資訊傳播出去，讓有興趣的眾人都受益。希望這個拋磚引玉的舉動能讓更多人響應與跟進，一起共襄盛舉。

　　本書可能還有不盡完美之處，非常歡迎您的指教與建議。近期還將推出其他 Arduino 相關應用與實作的書籍，敬請期待。

　　最後，請您立刻行動翻書閱讀。

蔡英德 於台中沙鹿靜宜大學主顧樓

目 錄

Maker 系列

本書是『Arduino 程式教學』的第七本書，主要是給讀者熟悉 Arduino 的對外說話模組：語音模組。Arduino 開發板最強大的不只是它的簡單易學的開發工具，最強大的是它封富的周邊模組與簡單易學的模組函式庫，幾乎 Maker 想到的東西，都有廠商或 Maker 開發它的周邊模組，透過這些周邊模組，Maker 可以輕易的將想要完成的東西用堆積木的方式快速建立,而且最強大的是這些周邊模組都有對應的函式庫，讓 Maker 不需要具有深厚的電子、電機與電路能力，就可以輕易駕御這些模組。

所以本書要介紹市面上最常見、最受歡迎與使用的顯示模組,讓讀者可以輕鬆學會這些常用模組的使用方法，進而提升各位 Maker 的實力。

本系列的新特色就是，筆者開始走小書系列，就是把獨立技能的部分獨立成專書，對於 Arduino 程式語法，請參考拙作『Arduino 程式教學(基本語法篇):Arduino Programming (Language & Syntax)』、『Arduino 程序教学(基本语法篇) :Arduino Programming (Language & Syntax)』(曹永忠, 許智誠, & 蔡英德, 2016a, 2016b)等書，本系列等書則不再詳述這些內容。

1

CHAPTER

揚聲器

由於許多電子線路必須將內部的狀態資訊，透過音效、語音等方式，發聲音到外部使用者，最常見的發聲音的簡單方式是使用簡單的揚聲器。

幸運的是，使用 Arduino 開發板，使用一個 Digital Pin(數位接腳)，連接到到處可見的揚聲器裝置(喇叭)，就可以透過 Arduino 開發板發出簡單，可辨識的音樂、特效、語音等，而且非常方便(曹永忠, 2016; 曹永忠, 許智誠, & 蔡英德, 2015a, 2015b, 2015c, 2015d; 曹永忠 et al., 2016a, 2016b)。

Tone 函式

使用 Arduino 開發板，使用一個 Digital Pin(數位接腳)連接喇叭，如本例子是接在數位接腳 13(Digital Pin 13)，讀者也可將喇叭接在您想要的腳位，只要將下列程式作對應修改，可以產生想要的音調。

我們將下表程式，請讀者鍵入Ｓｋｅｔｃｈ ＩＤＥ軟體(軟體下載請到：https://www.arduino.cc/en/Main/Software)，編譯完成後上傳到開發版進行測試。

範例：

```
#include <Tone.h>

Tone tone1;

void setup()
{
   tone1.begin(13);
   tone1.play(NOTE_A4);
```

```
}

void loop()
{
}
```

　　讀者可以由下表所示，可以了解使用 Tone 函數時，發出哪一種聲音，就必須輸入哪種參數。

表 1 Tone 頻率表

常態變數	頻率(Frequency (Hz))
NOTE_B2	123
NOTE_C3	131
NOTE_CS3	139
NOTE_D3	147
NOTE_DS3	156
NOTE_E3	165
NOTE_F3	175
NOTE_FS3	185
NOTE_G3	196
NOTE_GS3	208
NOTE_A3	220
NOTE_AS3	233
NOTE_B3	247
NOTE_C4	262

常態變數	頻率(Frequency (Hz))
NOTE_CS4	277
NOTE_D4	294
NOTE_DS4	311
NOTE_E4	330
NOTE_F4	349
NOTE_FS4	370
NOTE_G4	392
NOTE_GS4	415
NOTE_A4	440
NOTE_AS4	466
NOTE_B4	494
NOTE_C5	523
NOTE_CS5	554
NOTE_D5	587
NOTE_DS5	622
NOTE_E5	659
NOTE_F5	698
NOTE_FS5	740
NOTE_G5	784
NOTE_GS5	831
NOTE_A5	880

常態變數	頻率(Frequency (Hz))
NOTE_AS5	932
NOTE_B5	988
NOTE_C6	1047
NOTE_CS6	1109
NOTE_D6	1175
NOTE_DS6	1245
NOTE_E6	1319
NOTE_F6	1397
NOTE_FS6	1480
NOTE_G6	1568
NOTE_GS6	1661
NOTE_A6	1760
NOTE_AS6	1865
NOTE_B6	1976
NOTE_C7	2093
NOTE_CS7	2217
NOTE_D7	2349
NOTE_DS7	2489
NOTE_E7	2637
NOTE_F7	2794
NOTE_FS7	2960

常態變數	頻率(Frequency (Hz))
NOTE_G7	3136
NOTE_GS7	3322
NOTE_A7	3520
NOTE_AS7	3729
NOTE_B7	3951
NOTE_C8	4186
NOTE_CS8	4435
NOTE_D8	4699
NOTE_DS8	4978

資料來源：

https://code.google.com/p/rogue-code/wiki/ToneLibraryDocumentation#Ugly_Details

讀者可以由下表所示，可以了解使用 Tone 函數時，發出哪一種聲音，就必須輸入哪種參數。

表 2 Tone 音階頻率對照表

音階	常態變數	頻率(Frequency (Hz))
低音 Do	NOTE_C4	262
低音 Re	NOTE_D4	294
低音 Mi	NOTE_E4	330
低音 Fa	NOTE_F4	349
低音 So	NOTE_G4	392

音階	常態變數	頻率(Frequency (Hz))
低音 La	NOTE_A4	440
低音 Si	NOTE_B4	494
中音 Do	NOTE_C5	523
中音 Re	NOTE_D5	587
中音 Mi	NOTE_E5	659
中音 Fa	NOTE_F5	698
中音 So	NOTE_G5	784
中音 La	NOTE_A5	880
中音 Si	NOTE_B5	988
高音 Do	NOTE_C6	1047
高音 Re	NOTE_D6	1175
高音 Mi	NOTE_E6	1319
高音 Fa	NOTE_F6	1397
高音 So	NOTE_G6	1568
高音 La	NOTE_A6	1760
高音 Si	NOTE_B6	1976
高高音 Do	NOTE_C7	2093

資料來源：

https://code.google.com/p/rogue-code/wiki/ToneLibraryDocumentation#Ugly_Details

讀者可以由下圖所示，可以使用 Tone 函數，其硬體如何組立，為了方便，通常會加入一個降壓電阻。

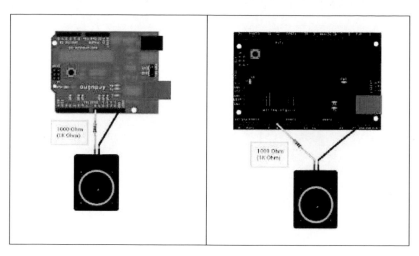

圖 1 Tone 接腳圖

資料來源：

https://code.google.com/p/rogue-code/wiki/ToneLibraryDocumentation#Ugly_Details

讀者可以由下圖所示，可以了解硬體組立之電路圖。

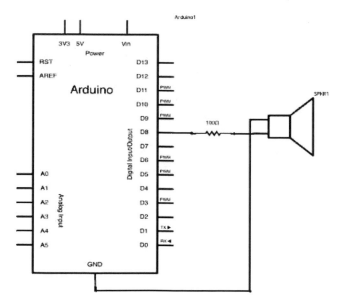

圖 2 Arduino 喇叭接線圖

如果要使用 Arduino 開發板發出一個完整的音樂，同樣使用一個 Digital Pin(數位接腳)連接喇叭，下列程式就可以產收馬力歐的音樂的音調。

Mario 音樂範例：

Mario 音樂範例(MarioBros)

```
/*
   Arduino Mario Bros Tunes
   With Piezo Buzzer and PWM
   by: Dipto Pratyaksa
   last updated: 31/3/13
*/
#include <pitches.h>

#define melodyPin 3
//Mario main theme melody
int melody[] = {
   NOTE_E7, NOTE_E7, 0, NOTE_E7,
   0, NOTE_C7, NOTE_E7, 0,
   NOTE_G7, 0, 0,   0,
   NOTE_G6, 0, 0, 0,

   NOTE_C7, 0, 0, NOTE_G6,
   0, 0, NOTE_E6, 0,
   0, NOTE_A6, 0, NOTE_B6,
   0, NOTE_AS6, NOTE_A6, 0,

   NOTE_G6, NOTE_E7, NOTE_G7,
   NOTE_A7, 0, NOTE_F7, NOTE_G7,
   0, NOTE_E7, 0,NOTE_C7,
   NOTE_D7, NOTE_B6, 0, 0,

   NOTE_C7, 0, 0, NOTE_G6,
   0, 0, NOTE_E6, 0,
   0, NOTE_A6, 0, NOTE_B6,
   0, NOTE_AS6, NOTE_A6, 0,
```

```
    NOTE_G6, NOTE_E7, NOTE_G7,
    NOTE_A7, 0, NOTE_F7, NOTE_G7,
    0, NOTE_E7, 0,NOTE_C7,
    NOTE_D7, NOTE_B6, 0, 0
};
//Mario main them tempo
int tempo[] = {
    12, 12, 12, 12,
    12, 12, 12, 12,
    12, 12, 12, 12,
    12, 12, 12, 12,

    12, 12, 12, 12,
    12, 12, 12, 12,
    12, 12, 12, 12,
    12, 12, 12, 12,

    9, 9, 9,
    12, 12, 12, 12,
    12, 12, 12, 12,
    12, 12, 12, 12,

    12, 12, 12, 12,
    12, 12, 12, 12,
    12, 12, 12, 12,
    12, 12, 12, 12,

    9, 9, 9,
    12, 12, 12, 12,
    12, 12, 12, 12,
    12, 12, 12, 12,
};

//

//Underworld melody
int underworld_melody[] = {
    NOTE_C4, NOTE_C5, NOTE_A3, NOTE_A4,
    NOTE_AS3, NOTE_AS4, 0,
```

```
   0,
   NOTE_C4, NOTE_C5, NOTE_A3, NOTE_A4,
   NOTE_AS3, NOTE_AS4, 0,
   0,
   NOTE_F3, NOTE_F4, NOTE_D3, NOTE_D4,
   NOTE_DS3, NOTE_DS4, 0,
   0,
   NOTE_F3, NOTE_F4, NOTE_D3, NOTE_D4,
   NOTE_DS3, NOTE_DS4, 0,
   0, NOTE_DS4, NOTE_CS4, NOTE_D4,
   NOTE_CS4, NOTE_DS4,
   NOTE_DS4, NOTE_GS3,
   NOTE_G3, NOTE_CS4,
   NOTE_C4, NOTE_FS4,NOTE_F4, NOTE_E3, NOTE_AS4, NOTE_A4,
   NOTE_GS4, NOTE_DS4, NOTE_B3,
   NOTE_AS3, NOTE_A3, NOTE_GS3,
   0, 0, 0
};
//Underwolrd tempo
int underworld_tempo[] = {
   12, 12, 12, 12,
   12, 12, 6,
   3,
   12, 12, 12, 12,
   12, 12, 6,
   3,
   12, 12, 12, 12,
   12, 12, 6,
   3,
   12, 12, 12, 12,
   12, 12, 6,
   6, 18, 18, 18,
   6, 6,
   6, 6,
   6, 6,
   18, 18, 18,18, 18, 18,
   10, 10, 10,
   10, 10, 10,
   3, 3, 3
```

```
};

void setup(void)
{
    pinMode(3, OUTPUT);//buzzer
    pinMode(13, OUTPUT);//led indicator when singing a note

}
void loop()
{
//sing the tunes
    sing(1);
    sing(1);
    sing(2);
}
int song = 0;

void sing(int s){
    // iterate over the notes of the melody:
    song = s;
    if(song==2){
        Serial.println(" 'Underworld Theme'");
        int size = sizeof(underworld_melody) / sizeof(int);
        for (int thisNote = 0; thisNote < size; thisNote++) {

            // to calculate the note duration, take one second
            // divided by the note type.
            //e.g. quarter note = 1000 / 4, eighth note = 1000/8, etc.
            int noteDuration = 1000/underworld_tempo[thisNote];

            buzz(melodyPin, underworld_melody[thisNote],noteDuration);

            // to distinguish the notes, set a minimum time between them.
            // the note's duration + 30% seems to work well:
            int pauseBetweenNotes = noteDuration * 1.30;
            delay(pauseBetweenNotes);

            // stop the tone playing:
            buzz(melodyPin, 0,noteDuration);
```

```
      }

    }else{

      Serial.println(" 'Mario Theme'");
      int size = sizeof(melody) / sizeof(int);
      for (int thisNote = 0; thisNote < size; thisNote++) {

        // to calculate the note duration, take one second
        // divided by the note type.
        //e.g. quarter note = 1000 / 4, eighth note = 1000/8, etc.
        int noteDuration = 1000/tempo[thisNote];

        buzz(melodyPin, melody[thisNote],noteDuration);

        // to distinguish the notes, set a minimum time between them.
        // the note's duration + 30% seems to work well:
        int pauseBetweenNotes = noteDuration * 1.30;
        delay(pauseBetweenNotes);

        // stop the tone playing:
        buzz(melodyPin, 0,noteDuration);

      }
    }
}

void buzz(int targetPin, long frequency, long length) {
   digitalWrite(13,HIGH);
   long delayValue = 1000000/frequency/2; // calculate the delay value
between transitions
   //// 1 second's worth of microseconds, divided by the frequency, then split
in half since
   //// there are two phases to each cycle
   long numCycles = frequency * length/ 1000; // calculate the number of
cycles for proper timing
   //// multiply frequency, which is really cycles per second, by the number of
seconds to
```

```
//// get the total number of cycles to produce
for (long i=0; i < numCycles; i++){ // for the calculated length of time...
    digitalWrite(targetPin,HIGH); // write the buzzer pin high to push out the
diaphram
    delayMicroseconds(delayValue); // wait for the calculated delay value
    digitalWrite(targetPin,LOW); // write the buzzer pin low to pull back the
diaphram
    delayMicroseconds(delayValue); // wait again or the calculated delay
value
}
digitalWrite(13,LOW);

}
```

Mario 音樂範例(pitches.h)

```
/*************************************************
 * Public Constants
 *************************************************/


#define NOTE_B0   31
#define NOTE_C1   33
#define NOTE_CS1 35
#define NOTE_D1   37
#define NOTE_DS1 39
#define NOTE_E1   41
#define NOTE_F1   44
#define NOTE_FS1 46
#define NOTE_G1   49
#define NOTE_GS1 52
#define NOTE_A1   55
#define NOTE_AS1 58
#define NOTE_B1   62
#define NOTE_C2   65
#define NOTE_CS2 69
#define NOTE_D2   73
#define NOTE_DS2 78
#define NOTE_E2   82
#define NOTE_F2   87
#define NOTE_FS2 93
```

```
#define NOTE_G2    98
#define NOTE_GS2 104
#define NOTE_A2    110
#define NOTE_AS2 117
#define NOTE_B2    123
#define NOTE_C3    131
#define NOTE_CS3 139
#define NOTE_D3    147
#define NOTE_DS3 156
#define NOTE_E3    165
#define NOTE_F3    175
#define NOTE_FS3 185
#define NOTE_G3    196
#define NOTE_GS3 208
#define NOTE_A3    220
#define NOTE_AS3 233
#define NOTE_B3    247
#define NOTE_C4    262
#define NOTE_CS4 277
#define NOTE_D4    294
#define NOTE_DS4 311
#define NOTE_E4    330
#define NOTE_F4    349
#define NOTE_FS4 370
#define NOTE_G4    392
#define NOTE_GS4 415
#define NOTE_A4    440
#define NOTE_AS4 466
#define NOTE_B4    494
#define NOTE_C5    523
#define NOTE_CS5 554
#define NOTE_D5    587
#define NOTE_DS5 622
#define NOTE_E5    659
#define NOTE_F5    698
#define NOTE_FS5 740
#define NOTE_G5    784
#define NOTE_GS5 831
#define NOTE_A5    880
```

```
#define NOTE_AS5 932
#define NOTE_B5   988
#define NOTE_C6   1047
#define NOTE_CS6 1109
#define NOTE_D6   1175
#define NOTE_DS6 1245
#define NOTE_E6   1319
#define NOTE_F6   1397
#define NOTE_FS6 1480
#define NOTE_G6   1568
#define NOTE_GS6 1661
#define NOTE_A6   1760
#define NOTE_AS6 1865
#define NOTE_B6   1976
#define NOTE_C7   2093
#define NOTE_CS7 2217
#define NOTE_D7   2349
#define NOTE_DS7 2489
#define NOTE_E7   2637
#define NOTE_F7   2794
#define NOTE_FS7 2960
#define NOTE_G7   3136
#define NOTE_GS7 3322
#define NOTE_A7   3520
#define NOTE_AS7 3729
#define NOTE_B7   3951
#define NOTE_C8   4186
#define NOTE_CS8 4435
#define NOTE_D8   4699
#define NOTE_DS8 4978
```

程式下載網址：https://github.com/brucetsao/eSound/

讓 Arduino 發出聲音

本實驗除了一塊 Arduino 開發板與 USB 下載線之外，我們加入小喇叭(如下圖.(c)所示)。

如下圖所示，這個實驗我們需要用到的實驗硬體有下圖.(a)的 Arduino Mega 2560、下圖.(b) USB 下載線、下圖.(c) 小喇叭。

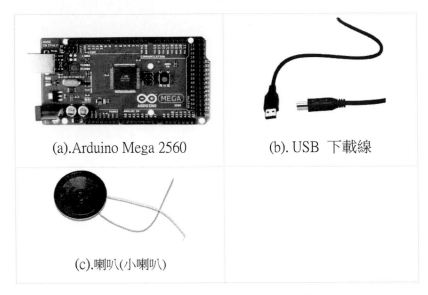

(a).Arduino Mega 2560	(b). USB　下載線
(c).喇叭(小喇叭)	

圖 3 讓 Arduino 發出聲音所需材料表

我們遵照前幾章所述，將 Arduino 開發板的驅動程式安裝好之後，遵照下圖之電路圖進行組裝。

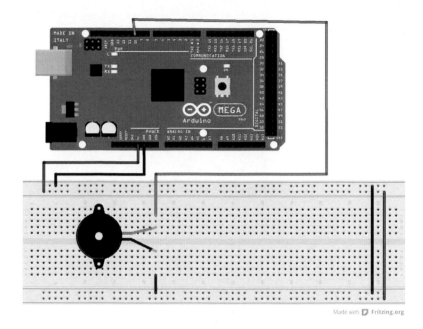

圖 4 讓 Arduino 發出聲音線路圖

我們將下表程式,請讀者鍵入Ｓｋｅｔｃｈ ＩＤＥ軟體(軟體下載請到:
https://www.arduino.cc/en/Main/Software),編譯完成後上傳到開發版進行測試。

我們發現 Arduino 開發板已經可以驅動小喇叭讓它發出歐揖、歐揖的聲音。

表 3 讓 Arduino 發出聲音程式

讓 Arduino 發出聲音程式(buzzer)

int buzzer=11;//設置控制蜂鳴器的數位 IO 腳
void setup()
{
pinMode(buzzer,OUTPUT);//設置數位 IO 腳模式,OUTPUT 為輸出
}
void loop()
{
unsigned char i,j;//定義變數
while(1)
{
for(i=0;i<80;i++)//輸出一個頻率的聲音
{

```
digitalWrite(buzzer,HIGH);//發聲音
delay(1);//延時 1ms
digitalWrite(buzzer,LOW);//不發聲音
delay(1);//延時 ms
}
for(i=0;i<100;i++)//輸出另一個頻率乩聲音
{
digitalWrite(buzzer,HIGH);//發聲音
delay(2);//延時 2ms
digitalWrite(buzzer,LOW);//不發聲音
delay(2);//延時 2ms
}
}
}
```

讀者也可以在筆者 YouTube 頻道(https://www.youtube.com/user/UltimaBruce)中，
在網址 https://www.youtube.com/watch?v=SQ3-6hn7jCg&feature=youtu.be，看到本次實
驗-讓 Arduino 發出聲音執行情形。

當然、如下圖所示，我們可以看到組立好的實驗圖，Arduino 開發板可以發出
聲音。

圖 5 讓 Arduino 發出聲音結果畫面

讓 Arduino 發出簡單音樂

前面筆者已經讓 Arduino 開發板發出聲音，但如果我們要讓 Arduino 開發板發出簡單音樂，我們必需加入新的指令：tone()，對於這個指令不熟的讀者，可以參考前幾章或是拙作『Arduino RFID 門禁管制機設計: The Design of an Entry Access Control Device based on RFID Technology』(曹永忠, 許智誠, & 蔡英德, 2014d)、『Arduino EM-RFID 門禁管制機設計:The Design of an Entry Access Control Device based on EM-RFID Card』(曹永忠, 許智誠, & 蔡英德, 2014b)、『Arduino RFID 门禁管制机设计: Using Arduino to Develop an Entry Access Control Device with RFID Tags 』(曹永忠, 許智誠, & 蔡英德, 2014c)、『Arduino EM-RFID 门禁管制机设计:Using Arduino to Develop an Entry Access Control Device with EM-RFID Tags』(曹永忠, 許智誠, & 蔡英德, 2014a)、『Arduino 互動跳舞兔設計: The Interaction Design of a Dancing Rabbit by Arduino Technology』(曹永忠, 許智誠, & 蔡英德, 2014e)、『Arduino 互动跳舞兔设计: Using Arduino to Develop a Dancing Rabbit with An Android Apps』(曹永忠, 許智誠, & 蔡英德, 2014a)，有興趣讀者可到 Google Books (https://play.google.com/store/books/author?id=曹永忠) & Google Play (https://play.google.com/store/books/author?id=曹永忠) 或 Pubu 電子書城 (http://www.pubu.com.tw/store/ultima) 購買該書閱讀之。。

本實驗除了一塊 Arduino 開發板與 USB 下載線之外，我們加入小喇叭(如圖 6.(c)所示)。

如圖 3 所示，這個實驗我們需要用到的實驗硬體有圖 6.(a)的 Arduino Mega 2560、圖 6.(b) USB 下載線、圖 6.(c) 小喇叭。

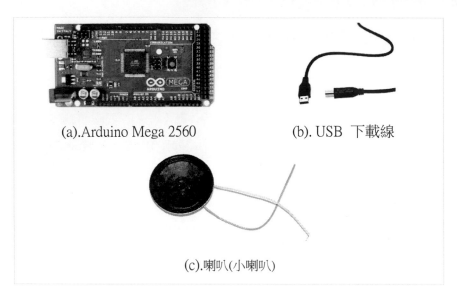

(a).Arduino Mega 2560　　　　　(b). USB 下載線

(c).喇叭(小喇叭)

圖 6 讓 Arduino 發出簡單音樂所需材料表

我們遵照前幾章所述，將 Arduino 開發板的驅動程式安裝好之後，遵照圖 7 之電路圖進行組裝。

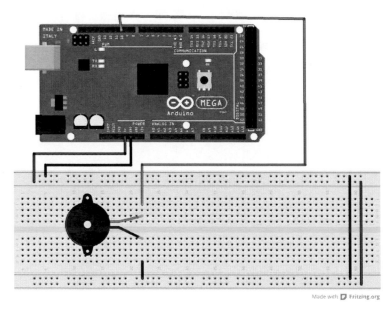

圖 7 讓 Arduino 發出簡單音樂線路圖

我們將下表程式，請讀者鍵入Ｓｋｅｔｃｈ　ＩＤＥ軟體(軟體下載請到：https://www.arduino.cc/en/Main/Software)，編譯完成後上傳到開發版進行測試。

我們發現Arduino開發板已經可以驅動小喇叭讓它發出瑪琍歐的簡單音樂。

表 4 讓 Arduino 發出簡單音樂程式

讓 Arduino 發出簡單音樂程式(toneMusic)
| ```
#include "pitches.h"

int melodyPin=11;//設置控制蜂鳴器的數位 IO 腳
int melody[] = {
    NOTE_E7, NOTE_E7, 0, NOTE_E7,
    0, NOTE_C7, NOTE_E7, 0,
    NOTE_G7, 0, 0,   0,
    NOTE_G6, 0, 0, 0,

    NOTE_C7, 0, 0, NOTE_G6,
    0, 0, NOTE_E6, 0,
    0, NOTE_A6, 0, NOTE_B6,
    0, NOTE_AS6, NOTE_A6, 0,

    NOTE_G6, NOTE_E7, NOTE_G7,
    NOTE_A7, 0, NOTE_F7, NOTE_G7,
    0, NOTE_E7, 0,NOTE_C7,
    NOTE_D7, NOTE_B6, 0, 0,

    NOTE_C7, 0, 0, NOTE_G6,
    0, 0, NOTE_E6, 0,
    0, NOTE_A6, 0, NOTE_B6,
    0, NOTE_AS6, NOTE_A6, 0,

    NOTE_G6, NOTE_E7, NOTE_G7,
    NOTE_A7, 0, NOTE_F7, NOTE_G7,
    0, NOTE_E7, 0,NOTE_C7,
    NOTE_D7, NOTE_B6, 0, 0
};

//Mario main them tempo
``` |
```

```
int tempo[] = {
  12, 12, 12, 12,
  12, 12, 12, 12,
  12, 12, 12, 12,
  12, 12, 12, 12,

  12, 12, 12, 12,
  12, 12, 12, 12,
  12, 12, 12, 12,
  12, 12, 12, 12,

  9, 9, 9,
  12, 12, 12, 12,
  12, 12, 12, 12,
  12, 12, 12, 12,

  12, 12, 12, 12,
  12, 12, 12, 12,
  12, 12, 12, 12,
  12, 12, 12, 12,

  9, 9, 9,
  12, 12, 12, 12,
  12, 12, 12, 12,
  12, 12, 12, 12,
};


  int mariolen = sizeof(melody) / sizeof(int) ;


void setup()
{
pinMode(melodyPin,OUTPUT);//設置數位 IO 腳模式，OUTPUT 為輸出
}
void loop()
{
    playMario() ;
}
```

```
void playMario()
{
    int noteDuration ;
    for(int mariopos=0; mariopos <mariolen; mariopos++)
    {
            noteDuration = 1000/tempo[mariopos];
            tone(melodyPin, melody[mariopos],noteDuration);
    }


}
```

表 5 讓 Arduino 發出簡單音樂程式之 h 檔

讓 Arduino 發出簡單音樂程式之 h 檔(pitches.h)
/***
* Public Constants
***/
#define NOTE_B0   31
#define NOTE_C1   33
#define NOTE_CS1 35
#define NOTE_D1   37
#define NOTE_DS1 39
#define NOTE_E1   41
#define NOTE_F1   44
#define NOTE_FS1 46
#define NOTE_G1   49
#define NOTE_GS1 52
#define NOTE_A1   55
#define NOTE_AS1 58
#define NOTE_B1   62
#define NOTE_C2   65
#define NOTE_CS2 69
#define NOTE_D2   73
#define NOTE_DS2 78
#define NOTE_E2   82

```
#define NOTE_F2    87
#define NOTE_FS2 93
#define NOTE_G2    98
#define NOTE_GS2 104
#define NOTE_A2    110
#define NOTE_AS2 117
#define NOTE_B2    123
#define NOTE_C3    131
#define NOTE_CS3 139
#define NOTE_D3    147
#define NOTE_DS3 156
#define NOTE_E3    165
#define NOTE_F3    175
#define NOTE_FS3 185
#define NOTE_G3    196
#define NOTE_GS3 208
#define NOTE_A3    220
#define NOTE_AS3 233
#define NOTE_B3    247
#define NOTE_C4    262
#define NOTE_CS4 277
#define NOTE_D4    294
#define NOTE_DS4 311
#define NOTE_E4    330
#define NOTE_F4    349
#define NOTE_FS4 370
#define NOTE_G4    392
#define NOTE_GS4 415
#define NOTE_A4    440
#define NOTE_AS4 466
#define NOTE_B4    494
#define NOTE_C5    523
#define NOTE_CS5 554
#define NOTE_D5    587
#define NOTE_DS5 622
#define NOTE_E5    659
#define NOTE_F5    698
#define NOTE_FS5 740
#define NOTE_G5    784
```

```
#define NOTE_GS5 831
#define NOTE_A5    880
#define NOTE_AS5 932
#define NOTE_B5    988
#define NOTE_C6    1047
#define NOTE_CS6 1109
#define NOTE_D6    1175
#define NOTE_DS6 1245
#define NOTE_E6    1319
#define NOTE_F6    1397
#define NOTE_FS6 1480
#define NOTE_G6    1568
#define NOTE_GS6 1661
#define NOTE_A6    1760
#define NOTE_AS6 1865
#define NOTE_B6    1976
#define NOTE_C7    2093
#define NOTE_CS7 2217
#define NOTE_D7    2349
#define NOTE_DS7 2489
#define NOTE_E7    2637
#define NOTE_F7    2794
#define NOTE_FS7 2960
#define NOTE_G7    3136
#define NOTE_GS7 3322
#dcfinc NOTE_A7    3520
#define NOTE_AS7 3729
#define NOTE_B7    3951
#define NOTE_C8    4186
#define NOTE_CS8 4435
#define NOTE_D8    4699
#define NOTE_DS8 4978
```

　　讀者也可以在筆者 YouTube 頻道

(https://www.youtube.com/user/UltimaBruce)中，在網址

https://www.youtube.com/watch?v=kgQMVnpz6YM&feature=youtu.be，看到本次實驗-讓

Arduino 發出瑪琍歐的簡單音樂執行情形。

當然、如下圖所示，我們可以看到組立好的實驗圖，Arduino 開發板可以發出瑪琍歐的簡單音樂。

圖 8 讓 Arduino 發出簡單音樂結果畫面

章節小結

本章節介紹到處可見的揚聲器裝置(喇叭)，就可以透過 Arduino 開發板發出簡單，可辨識的音樂、特效、語音等，透過以上章節的內容，一定可以一步一步的將內容給予實作出來。

2

CHAPTER

語音基本介紹

何謂 MP3

MP3 是一種壓縮檔案的格式,用來儲存數位的聲音資料。 一般說來,透過電腦使用 MP3 格式壓縮之後的資料量,大約只有原本的 10 到 12 分之一。但是這樣的壓縮方式會使原本的資料失去一部份,壓的愈小,失真也會愈多,所以稱為失真壓縮[1]。

MP3 原文是 ISO-MPEG Audio Layer-3(全球影像/聲音/系統壓縮標準第三級)的簡稱,它是在 1987 年的 Digital AudioBroadcasting[2]計畫中所開發出的音樂聲音壓縮方法,但 MP3 與 MPEG-3 是完全不一樣的東西。它是一種聲頻壓縮格式。這種聲頻壓縮格式可以容許較高的壓縮(檔案小),但不減損聲音的高質量。MP3 主要的觀念是依靠一種感知編碼(Perceptual Coding)[3]的方法,因為透過感知編碼的方法,

---

[1] 資料壓縮一般說來,有「無失真壓縮」和「失真壓縮」兩類資料壓縮技術,無失真壓縮就是將壓縮資料還原時,與原來壓縮前的資料無異;反之,失真壓縮就是將壓縮資料還原時,與原來壓縮前的資料有所差異.

[2] Digital Audio Broadcasting (DAB) is a digital radio technology for broadcasting radio stations, used in several countries, particularly in Europe. As of 2006, approximately 1,002 stations worldwide broadcast in the DAB format.

The DAB standard was initiated as a European research project in the 1980s. The Norwegian Broadcasting Corporation (NRK) launched the very first DAB channel in the world on 1 June 1995 (NRK Klassisk), and the BBC and SR launched their first DAB digital radio broadcasts in September 1995. DAB receivers have been available in many countries since the end of the 1990s. DAB may offer more radio programmes over a specific spectrum than analogue FM radio. DAB is more robust with regard to noise and multipath fading for mobile listening, since DAB reception quality first degrades rapidly when the signal strength falls below a critical threshold, whereas FM reception quality degrades slowly with the decreasing signal.

[3] 一般來說,資料壓縮有兩種方法。一種方法是利用信號的統計性質,完全不丟失資訊的高效率編碼法,稱為平均信息量編碼或熵編碼。第二種方法是利用接收信號的人的感覺特性,省略不必要的資訊,壓縮信息量,這種方法稱為感覺編碼。

可以讓我們獲得獲得較小的檔案大小，同時又保證聲音的高質量，而人們可以感受到的相近似的聲音保真度。

因為 MP3 音樂檔案是經過壓縮之後的結果，所以它的聲音當然會與原本的有所差距，只不過對於一般人來說是不太容易分辨出來的。這也是為什麼 MP3 音樂這麼受歡迎的原因，一方面檔案小，一方面聲音不會差，由於檔案體積小而且聲音保真度高，這也是 MP3 在網路上如此盛行的原因(曹永忠, 2016; 曹永忠, 許智誠, et al., 2014a; 曹永忠, 許智誠, & 蔡英德, 2014b; 曹永忠, 許智誠, et al., 2014e; 曹永忠, 許智誠, & 蔡英德, 2014f)。

聲音壓縮

一般說來，聲音資料數位化之後，其資料量是非常龐大的，所以我們引入了壓縮的概念：

對於沒有進行壓縮的聲音資料，一般稱為：無損音質[4]，其實就是真實的聲音原始呈現，如：我們說話的聲音，樂器演奏的聲音；但是，由於數位音樂，又可分為 CD 或 WAV 以及 MP3 等等，並且，由於 MP3 通常皆由 CD 轉檔而成，故我們將 CD、WAV 等這些格式稱為『無損音頻』。

另一種進行壓縮的聲音資料稱為：有損壓縮[5]，是指經過壓縮、解壓的數據與

---

[4] 為自由無損音頻壓縮編碼(Free Lossless Audio Codec),FLAC 是一款著名的自由音頻壓縮編碼，其特點是可以對音訊檔案無失真壓縮。不同於其他有損壓縮編碼如 MP3 及 WMA（9.0 版本支援無失真壓縮），它不會破壞任何原有的音頻資訊，所以可以還原音樂光碟音質。現在它已被很多軟體及硬體音頻產品所支援。(資料來源：http://zh.wikipedia.org/wiki/FLAC)

[5] 有損壓縮是對利用了人類是絕對圖像或聲波中的某些頻率成分不敏感的特性，允許壓縮過程中損失一定的信息；雖然不能完全回復原始數據，但是所損失的部分對理解原始圖像的影響縮小有損壓縮 ，卻換來了大得多的壓縮比。有損壓縮廣泛應用於語音，圖像和視頻數據的壓縮。(資料來源：
http://www.twwiki.com/wiki/%E6%9C%89%E6%90%8D%E5%A3%93%E7%B8%AE)

原始數據不同但是非常接近的壓縮方法，或稱為『有損格式』。常見的有損格式中有 MP3 檔、RM 檔…等等，簡單來說，MP3 就是將 CD 音樂中人耳無法聽到的聲音範圍刪除，進而減少檔案大小，但在此時，原本的音質也被破壞了。

上述所敘之無損音頻，WAV 就是一種最常用的無損音頻檔案格式，由微軟 (Microsoft)公司開發。WAV 是一種檔案格式，符合 RIFF (Resource Interchange File Format)[6]規範。

所有的 WAV 都有一個檔案頭，這個檔案頭儲存音頻流的編碼參數。WAV 對『音頻流』的編碼沒有硬性規定，除了脈衝編碼調變(Pulse-code modulation：PCM) [7]之外，還有幾乎所有支援 PCM 規範的編碼都可以為 WAV 的音頻流進行編碼。

WAV 的介紹

WAV 數位格式檔案是最常且方便使用之數位音樂檔案，它最大的特色是未經任何壓縮處理，因此能表現最佳的聲音品質，同時無論是 PC 或是 MAC 之操作系統都能使用，當然也很容易使用於音源取樣器。

---

[6]資源交換檔案標準（Resource Interchange File Format，RIFF）是一種把資料儲存在被標記的區塊（tagged chunks）中的檔案格式（meta-format）。它是在 1991 年時，由 Microsoft 和 IBM 提出。它是 Electronic Arts 在 1985 提出的 Interchange File Format 的翻版。這兩種標準的唯一不同處是在多位元整數的儲存方式。 RIFF 使用的是小端序，這是 IBM PC 使用的處理器 80x86 所使用的格式，而 IFF 儲存整數的方式是使用大端序，這是 Amiga 和 Apple Macintosh 電腦使用的處理器，68k，可處理的整數型態。Microsoft 在 AVI 和 WAV 這兩種著名的檔案格式中，都使用 RIFF 的格式當成它們的基礎。(資料來源：http://zh.wikipedia.org/wiki/Resource_Interchange_File_Format)

[7] 是一種類比訊號的數位化方法。PCM 將訊號的強度依照同樣的間距分成數段，然後用獨特的數位記號（通常是二進位）來量化。(資料來源：
http://zh.wikipedia.org/wiki/%E8%84%88%E8%A1%9D%E7%B7%A8%E7%A2%BC%E8%AA%BF%E8%AE%8A)

相較於傳統的類比式音頻素材 CD (Audio CD)[8]，不用還要辛苦透過取樣器將聲音取樣進來並修剪前後音訊，數位 WAV 格式可以直接使用在作品中，透過音樂編輯軟體，可以使用滑鼠輕易拖拉成循環樂段(loops)，節省大量的後製編輯時間。當然也不必擔心檔案相容性的問題，因為數位 WAV 檔案為通用的音樂格式，所有的音樂編輯軟體都相容此音樂格式。

WAV 的優缺點:

- 優點：未經任何壓縮處理，因此能表現最佳的聲音品質，原音重現、品質高所有的音樂編輯軟體都相容此音樂格式。

- 缺點：檔案太大，對存儲空間需求太大，至使不便於網路交流和傳播，一般而言，使用 Wav 儲存的聲音資料，1 分鐘需要約 10 MB 以少， 對於低容量的儲存限制或應用於低網路頻寬而言，這可能是一個重要的議題。

---

[8] Super Audio CD（SACD）是由 Sony 及飛利浦兩家公司於 1999 年所訂定的音源儲存媒體，SACD 是 Sony 與 Philips 合力研發的音樂碟片規格，是繼 CD 的發明之後，成功超越 CD 錄音品質的新產品。音樂 CD（CDDA）採取 44.1KHz 的取樣頻率，因此在接近人耳聽覺極限的高頻訊號中（接近 20KHz）只能取樣出三點；要用這三點來恢復正確的類比訊號有困難性，被認為會造成相位誤差，人類的聽覺系統可以輕易的察覺相位失真。SACD 的取樣頻率高達 2822.4kHz，是一般 CD 44.1KHz 取樣的 64 倍，而且 SACD 頻率範圍更是高達 100KHz 以上；也因此使得 SACD 改善了原來音樂 CD 音質給人冷硬的刻板印象，而以更細膩、更多細節、更柔軟的聲音呈現。SACD 的錄音方式是用 Direct Stream Digital（DSD，即直流數位技術）方式錄音，摒除傳統的 PCM 錄音方式，將所有訊號以每秒 280 萬次直接把類比音樂訊號波形轉變為數位訊號，也就是所謂的『直接位元流數位』，因此取樣波形非常接近原來的類比波形。另外，SACD 省去位元轉換程式，降低了數位濾波而可能產生的失真與雜訊。還有一個特點就是 SACD 也可以容納多聲道以及影像，由於 SACD 自身的定位以及 1bit 量化 DSD 直接資料流程在技術方面的簡潔和優勢（多數 DAC 是處理 DSD 數位訊號及類比訊號的互相轉換，如果要輸出或輸入 PCM 格式，則必須加上 DSD 及 PCM 訊號的轉換機制，這個機制需要相當的計算量），使得大多數的資深音響發燒友經過親耳聆聽後，主觀感覺都認為 SACD 在音質上勝過音樂 CD。(資料來源：http://zh.wikipedia.org/wiki/Super_Audio_CD)

我們將 WAV 和 MP3 作一個簡單的比較表,列於表 6 之中:

表 6 WAV 和 MP3 的比較表

資料格式	簡介	優點	缺點
WAV 檔	聲音檔案的原始格式,主要是從 CD 或錄音帶直接轉換而來。	原音呈現,格式簡單與標準化	資料空間龐大,傳播不易。
MP3 檔	聲音檔的壓縮格式之一	壓縮率高(約為原來的 12 分之一)	某種程度的失真(人耳不易發覺)

位元率:

提到 Wav 與 MP3,我們就必需提到位元率,相信有許多人並不了解何謂位元率[9],在這邊簡單的說明一下位元率為何:

『位元率』簡單來說,就是每秒拿多少資料來表現這一秒的聲音,舉個例子來說:我們就以 WAV 與 MP3 來做一個比較與說明,WAV 就如同 DVD 影片,畫質較好,但檔案大小過大,而 MP3 就如同 VCD,雖然沒有 DVD 的畫質,但是,檔案較小。

方程式 1 位元率計算公式

$$位元率(\text{ Bit Rate})=(\text{Hz 取樣率})*(聲道數)*(位元深度\ \text{Bit})$$

由以上所示,我們可以看到 WAV 檔所使用的位元率與其它相關檔案的格式;而 WAV 檔案之容量是如何而來的呢?

---

[9]位元率(Bit Rate)是單位時間內傳輸送或處理的位元的數量。 位元率 規定使用「位元每秒」(bit/s 或 bps)為單位

WAV 檔之計算方式:

計算公式為位元率除以 8 等於此檔案每秒所使用的磁碟空間（KB/S），此時，再將剛剛計算出來的數值乘以一首歌的秒數後，即可得到檔案之容量。

方程式 2 WAV 檔案容量計算公式

$$WAV 檔案大小(bytes) = (Hz 取樣率) * (聲道數) * (位元深度 \ Bit)/8$$

以一首五分鐘的歌為例：計算 WAV 音樂的檔案大小

- 1411.2 Kbps / 8 (bits) =176.4 KB/s （單位換算）

- 176.4 KB/s * (60 秒 * 5 分鐘) = 52920 KB （檔案大小）

- 52920 KB / 1024 (bytes) = 約等於 51.68 MB

計算 MP3 音樂的檔案大小與 WAV 檔之計算方法相同。

以一首五分鐘的音樂為例：

- 192 Kbps / 8 (bits) = 24 KB/s （單位換算）

- 24 KB/s * (60 秒 * 5 分鐘) = 7200 KB（檔案大小）

- 7200 / 1024 (bytes) = 約等於 7 MB

章節小結

本章節主要介紹讀者聲音的基本原理，後面內容對於讀者播放聲音，有更深入的體認。

3

CHAPTER

WT588D-U 語音模組

本章節要介紹，在開發板常常使用的語音模組，WT588D-U 語音模組是一塊簡單、好用，音質佳的語音模組，也是許多開發者最佳選擇。

如何使用 WT588D-U 語音模組

由於解譯 Wav 或 Mp3 聲音資料，對於 CPU 而言，會耗損大量的 CPU 執行時間，並且對於攥寫解譯 Wav 或 Mp3 聲音資料的程式也是很大的負擔，所以本書引入了解譯 Wav 或 Mp3 聲音資料的硬體模組：WT588D-U 語音模組，下列將 WT588D-U 語音模組產品特徵列舉如下：

- 28 腳模組封裝，可通過更換記憶體以獲得不同長度的語音存儲時間
- 支援 2M bit ～32M bit 容量的 SPI-Flash（注：1byte=8bit）
- 採用 WT588D-20SS 語音晶片當作主控核心
- 內嵌獨特的人聲語音處理器，使語音表現極為自然悅耳
- 內置 13Bit/DA 轉換器，以及 12Bit/PWM 音訊處理，確保高品質語音輸出
- 支援載入 6K～22KHz 取樣速率 WAV 音訊
- PWM 輸出可直接推動 0.5W/8Ω揚聲器
- 支援 DAC/PWM 兩種輸出方式；
- 支援按鍵控制模式、一線串口控制模式、三線串口控制模式；
- 按鍵控制模式底下可以設置多種 IO 口觸發方式；
- 任意設定顯示語音播放狀態信號的 BUSY 輸出方式；
- 最多可載入 500 段編輯的語音
- 220 段可控制位址位，每個位址位最多可載入 128 段語音，地址位元內的語音組合播放
- 語音播放停止馬上進入休眠模式

- 搭配套 WT588D VoiceChip 控制軟體，介面簡單，使用方便。能發揮 WT588D-U 語音模組各項功能

- 軟體中可完成控制模式設置、語音組合、調用語音、靜音等操作

- 可隨意進入靜音，靜音時間範圍為 10ms～25min

- USB 下載方式，支援線上下載/離線下載；即便是在 WT588D-U 語音模組通電的情冴下，也一樣可以正常下載資料到 SPI-Flash

- 工作電壓 DC2.8V～5.5V

- 休眠電流小亍 10uA

- 抗干擾性強，可應用在工業領域

下列將 WT588D-U 語音模組功能描述列舉如下：

- 按鍵控制模式觸發方式靈活，可隨意設置任意按鍵為脈衝

- 可隨意設置任意按鍵為脈衝重複觸發

- 可隨意設置任意按鍵為脈衝可重複觸發

- 可隨意設置任意按鍵為無效按鍵

- 可隨意設置任意按鍵為上一曲

- 可隨意設置任意按鍵為下一曲

- 可隨意設置任意按鍵為上一曲可迴圈

- 可隨意設置任意按鍵為下一曲可迴圈

- 可隨意設置任意按鍵為音量+

- 可隨意設置任意按鍵為音量-

- 可隨意設置任意按鍵為播放

- 可隨意設置任意按鍵為暫停

- 可隨意設置任意按鍵為停止

- 可隨意設置任意按鍵為播放/停止

- 觸發方式：

- 一線串口控制模式

- 三線串口控制模式

- 可透過單晶片控制語音播放、停止、迴圈播放和音量大小，戒者直接觸發 0～219 地址位的任意語音。

下列將 WT588D-U 語音模組應用範圍列舉如下：

- 報站器

- 報警器

- 提醒器

- 鬧鐘

- 學習機

- 智慧家電

- 電子玩具

- 電訊

- 倒車雷達

- 各種自動控制裝置

本書使用的 WT588D-U 語音模組，由下圖所示，其連接電路非常簡單，若讀者想要其它連接方法的電路圖，可以參附錄章節中：『 WT588D-U 語音模組(英文版)』、『 WT588D-語音模組(中文版)』等相關資料(曹永忠, 許智诚, et al., 2014a, 2014b; 曹永忠, 許智誠, et al., 2014e, 2014f)。

讀者可以到筆者 Github(https://github.com/)網站，本書的所有範例檔，都可以在 https://github.com/brucetsao/eRabbit，下載所需要的檔案，對於音效範例檔部份，可以在 https://github.com/brucetsao/eRabbit/tree/master/wav 下載所需要音效範例檔(曹永忠, 許智诚, et al., 2014a, 2014b; 曹永忠, 許智誠, et al., 2014e, 2014f)。

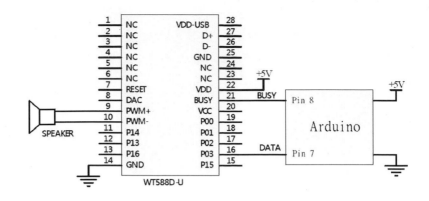

圖 9 WT588D-U 串列電路圖

首先，請讀者依照下表進行 WT588D-U 電路組立，再進行程式攥寫的動作。

表 7 WT588D-U 接腳表

	模組接腳	Arduino 開發板接腳	解說
WT588D-U	Pin 22	Arduino +5V	WT588D-U 資料接腳
	Pin 14	Arduino GND(共地接點)	
	Pin 21	接 LED 負級，正級連接 470R 到 Arduino +5V	
	Pin 9	外接喇叭 +Speaker	
	Pin 10	外接喇叭 -Speaker	
	Pin 16(P03)	Arduino digital output pin 7	

完成 Arduino 開發板與 playstation 搖桿連接之後，我們將下表程式，請讀者鍵入Ｓｋｅｔｃｈ　ＩＤＥ軟體(軟體下載請到：

https://www.arduino.cc/en/Main/Software)，編譯完成後上傳到開發版進行測試。

將下表之 WT588D-U 語音模組測試程式一上載到 Arduino 開發板進行測試，可以聽到 WT588D-U 語音模組發出聲音，因為 WT588D-U 語音模組推動喇叭的電力並非很強大，若聲音太小時，讀者可以外接音源擴大機將聲音放大。

表 8 WT588D-U 語音模組測試程式一

WT588D-U 語音模組測試程式一(WT588_t01)

```
int addrpin=7;

int busypin=8;

char addr;

void PlayVoice(unsigned char addr,unsigned int addrpin)

{

digitalWrite(addrpin,0);

delay(5);

for(int i=0;i<8;i++)

{

digitalWrite(addrpin,1);

if(addr & 1)

{

delayMicroseconds(600);

digitalWrite(addrpin,0);

delayMicroseconds(300);

}

else

{

delayMicroseconds(300);

digitalWrite(addrpin,0);

delayMicroseconds(600);

}

addr>>=1;   //此行用>>=还是=>>还不确定

}

digitalWrite(addrpin,1);
```

```
}




/*用于播放由一串地址组成的语句*/

void PlayVoiceSerial(unsigned int addrserial[],unsigned int len,unsigned int addrpin,unsigned int busypin)

{

for(int i=0;i<len;i++)

{

addr=addrserial[i];

PlayVoice(addr,addrpin);

pulseIn(busypin,0);

pulseIn(busypin,1);

}

}









void setup()

{


pinMode(addrpin, OUTPUT);
```

WT588D-U 語音模組測試程式一(WT588_t01)

```
}



void loop()

{

    delay(2000);

    addr=0;



    PlayVoice(addr,addrpin);

    pulseIn(busypin,0);

pulseIn(busypin,1);

delay(5000);

addr=1;

PlayVoice(addr,addrpin);

    pulseIn(busypin,0);

pulseIn(busypin,1);

}
```

透過外界參數使用 WT588D-U 語音模組

　　由於上述程式已經可以完整驅動 WT588D-U 語音模組，對於如何燒錄 Wav 語音檔案到 WT588D-U 語音模組，請讀者參考附錄中『WT588D 語音燒錄器操作手冊』一節，便可以輕易將 Wav 語音檔案燒錄到 WT588D-U 語音模組。

　　由於我們一段語音只有燒錄一個對應的 Wav 語音檔案，再針對該段語音，透過串列通訊來驅動該段語音，所以我們必需改寫程式，所以將下列之 W WT588D-U

語音模組測試程式二，鍵入 Arduino Sketch 之中(軟體下載請到：https://www.arduino.cc/en/Main/Software)，完成編譯後，上載到 Arduino 開發板進行測試，我們可以在 Arduino Sketch 監控畫面之中，透過輸入 0~9，聽到第一段到第十段的語音。

所以我們可以見到下圖所示，使用者可以在 Arduino Sketch 監控畫面之中，透過輸入 0~9，聽到第一段到第十段的語音。

表 9 WT588D-U 語音模組測試程式二

WT588D-U 語音模組測試程式二(WT588_t02)
| ```
int addrpin=7;
int busypin=8;
char addr;


void setup()
{
pinMode(addrpin, OUTPUT);
Serial.begin(9600);
Serial.println("program Start here");
}



void loop()
{
  int aa ;
  char song;
``` |
```

```
    if (Serial.available())
    {
        aa =   Serial.read() ;


        song = (char)aa -0x30;
            Serial.print("now playing song #");
        Serial.println(song,HEX) ;
    PlayVoice(song,addrpin);
    }
//;
}
void PlayVoice(unsigned char addr,unsigned int addrpin)
{
digitalWrite(addrpin,0);
delay(5);
for(int i=0;i<8;i++)
{
digitalWrite(addrpin,1);
if(addr & 1)
{
delayMicroseconds(600);
digitalWrite(addrpin,0);
delayMicroseconds(300);
}
else
```

```
{
delayMicroseconds(300);

digitalWrite(addrpin,0);

delayMicroseconds(600);

}

addr>>=1;   //此行用>>=还是=>>还不确定

}

digitalWrite(addrpin,1);

}



/*用于播放由一串地址组成的语句*/

void PlayVoiceSerial(unsigned int addrserial[],unsigned int len,unsigned int addrpin,unsigned int busypin)

{

for(int i=0;i<len;i++)

{

addr=addrserial[i];

PlayVoice(addr,addrpin);

pulseIn(busypin,0);

pulseIn(busypin,1);

}

}
```

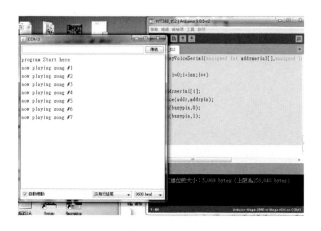

圖 10 WT588D-U 語音模組測試程式二執行畫面

音效檔轉檔

我們發現 WT588D-U 語音模組需要某些格式的 Wav 音樂檔，但是我們找到的 Wav 音樂檔無法符合 WT588D-U 語音模組的工具程式『wt588d voicechip』的規定，此時我們必須將檔 Wav、Mp3、Wma…等等音 Wav 音樂檔效檔進行轉檔。

筆者採用 FreeMake 公司的產品『Free Audio Converter』，進行轉檔，讀者可以到：http://www.freemake.com/tw/free_audio_converter/ 、 或 到 筆 者 Github ： https://github.com/brucetsao/eSound/tree/master/Tools，下載該軟體。

首先，讀者下載該軟體之後，透過標準的安裝方式，安裝好該軟體之後，如下圖所示，啟動該軟體。

圖 11 Free Audio Converter 啟始畫面

讀者可以透過檔案總管等軟體，將要轉檔的音效檔，拖拉到『Free Audio Converter』軟體畫面，如下圖所示，設定好要轉檔的音效檔。

圖 12 將要轉檔音效檔拖拉到畫面

如下圖所示讀者必需先透過行設定轉檔之取樣頻率、單或立體音。由於 WT588D-U 語音模組的工具程式『wt588d voicechip』的規定，注意!如果要將一般 mp3 檔案轉 wav 檔時，記得先將檔案設為單音(MONO)、取樣率設定為 6KHz、8KHz、12KHz、16KHz、22KHz，取樣大小設定為 16 bit。以上取樣率請皆設為整數，如取樣率為 22.001KHz 時，則軟體將會警告無法正常讀入，請讀者多加注意。

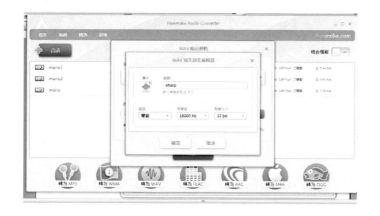

圖 13 設定轉檔取樣頻率等資料

如下圖所示，讀者必需先透過行設定轉檔之取樣頻率、單或立體音設定與轉出路徑後，按下下圖所示之『轉換』鈕，開始進行檔案格式轉換。

圖 14 設定正確取樣資料與轉出路徑

若一切就緒，按下上圖所示之『轉換』鈕之後，開始進行檔案格式轉換，如下圖所示，則完成音效檔的格式轉換後，讀者就可以依附錄之 WT588D 之相關資料，將音效檔燒入『WT588D-U 語音模組』，進行相關實驗了。

圖 15 音效檔格式轉換完成

章節小結

本章主要介紹之 Arduino 開發板使用與連接 WT588D-U 語音模組，透過本章節的解說，相信讀者會對連接、使用 WT588D-U 語音模組，有更深入的了解與體認。

4

CHAPTER

Arduino Wave Module V2 with 2G SD card

ELECHouse 出 廠 的 Arduino Wave Shield V2 （ 網 址 ：
http://www.elechouse.com/elechouse/index.php?main_page=product_info&cPath=168_
170&products_id=2161） ， 讀 者 可 以 從 台 灣 德 源 科 技 的 賣 場 ：
http://goods.ruten.com.tw/item/show?21307113175854，買到這篇音效擴充板。

Arduino Wave 模組

Arduino Wave Shield V2(網址：
http://www.elechouse.com/elechouse/index.php?main_page=product_info&cPath=168_170&
products_id=2161)(如下圖所示)，他支援 14KHz/16bit WAV 音效檔與 32KHz/16bit
AD4 音效檔，本張音效模組支援 DAC 輸出與 PWM 輸出，對於音效檔案，可以
使用 SD/MMC 記憶卡，本模組只需要三個腳位就可以控制音樂撥放，所以說是一
個非常方便的音效模組。

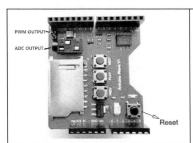

圖 16 ELECHOUSE WAVE V2.1 Shield

資料來源：ELECHOUSE 官網：

http://www.elechouse.com/elechouse/index.php?main_page=product_info&cPath=168_170&products_id=2161

下列為 Arduino Wave Shield V2 的規格資料：

- 支援 SD 卡擴充音效檔(最大 1GB)。

- 支援 AD4 格式之音效檔 (6KHZ ~ 36KHZ) 與 WAV 格式之音效檔(6KHz ~14KHz)。

- 支援 4 Bit ADCPM 格式之音效檔。

- 支援耳機插座與外接楊聲器。

- 支援 16 Bit DAC / PWM 音效輸出。

- 只需要三個腳位就可以控制。

- 待機電流: 3uA

- 單聲道輸出

電路組立

首先，組立 Arduino Wave Shield V2 非常簡單，只要把 Arduino Wave Shield V2 接在 Arduino UNO 開發板上方，如下圖所示，只要將 Arduino Wave Shield V2 擴展板堆疊在 Arduino UNO 開發板就完成電路組立。

圖 17 Arduino Wave Shield V2 接腳圖

或到筆者 Github：https://github.com/brucetsao/eSound/tree/master/SoundFiles，下載測試音效檔：0000.wav，並將該檔案存入 SD 記憶卡的根目錄之下。

如下圖所示，為 Arduino Wave Shield SD 卡插槽處，將安裝好音效檔的 SD 記憶卡插入下圖紅框處，該模組只可以支援到 1G 容量的 SD 記憶卡。

圖 18 Arduino Wave Shield SD 卡插槽處

我們請讀者鍵入Ｓｋｅｔｃｈ　ＩＤＥ軟體（軟體下載請到：https://www.arduino.cc/en/Main/Software），編譯完成後上傳到開發版進行測試。

表 10 Arduino Wave Shield 測試程式一

Arduino Wave Shield 測試程式一(WaveShield01)

```
/*
    This code is show how Arduino Wave Module works with Arduino.
    Code is not optimized. Any improving work on it is encouraged.
    (C) Copyright 2011 elechouse.com
    */
#define PlayCmd 0xfffe
#define TurnOffCmd 0xfff0
#define Play_PauseCmd 0xfffe




int RST = 3;
int CLK = 9;
int DAT = 8;


void setup() {


    pinMode(RST, OUTPUT);
    pinMode(CLK, OUTPUT);
    pinMode(DAT, OUTPUT);


    digitalWrite(RST, HIGH);
    digitalWrite(CLK, HIGH);
    digitalWrite(DAT, HIGH);


    digitalWrite(RST, LOW);
    delay(5);
    digitalWrite(RST, HIGH);
    delay(300);
}

void loop() {
    send(0xfff7);//set voice volumn to 7
```

```
  send(0x0000);//play file 0000
  delay(400000);//delay 10 seconds



  send(Play_PauseCmd);// pause
  delay(5000);
  send(Play_PauseCmd);//play



  while(1);
}


/**************************************************
The following function is used to send command to wave shield.
You don't have to change it.

Send the file name to play the audio.
If you need to play file 0005.AD4, write code: send(0x0005).
For more command code, please refer to the manual
**************************************************/
void send(int data)
{
  digitalWrite(CLK, LOW);
  delay(2);
  for (int i=15; i>=0; i--)
  {
    delayMicroseconds(50);
    if((data>>i)&0x0001 >0)
      {
        digitalWrite(DAT, HIGH);
        //Serial.print(1);
      }
    else
      {
        digitalWrite(DAT, LOW);
        // Serial.print(0);
      }
```

```
    delayMicroseconds(50);
    digitalWrite(CLK, HIGH);
    delayMicroseconds(50);

    if(i>0)
    digitalWrite(DAT, LOW);
    else
    digitalWrite(DAT, HIGH);
    delayMicroseconds(50);

    if(i>0)
    digitalWrite(CLK, LOW);
    else
    digitalWrite(CLK, HIGH);
  }

  delay(20);
}
```

程式下載網址：https://github.com/brucetsao/eSound/

再將程式上傳到開發板之後，我們可以聽到插入耳機插孔的喇叭，可以發出對應的音樂或音效。

章節小結

本章節介紹 ELECHouse 出廠的 Arduino Wave Shield V2，主要是讓讀者了解 Arduino 開發板如何透過 ELECHouse 出廠的 Arduino Wave Shield V2，讓 Arduino 開發板可以發出音效，透過以上章節的內容，一定可以一步一步的將音效輸出給予實作出來。

5

CHAPTER

Serial MP3 Player

Grove - Serial MP3 Player 是 Seeed Studio 出廠的 Grove - Serial MP3 Player V1(網址：http://www.seeedstudio.com/depot/grove-serial-mp3-player-p-1542.html)，讀者 可 以 從 Seeed Studio 的 賣 場： http://www.seeedstudio.com/depot/grove-serial-mp3-player-p-1542.html，買到這篇音效擴充板，另外 Seeed Studio 也有一片進階的版本，是 Grove - Serial MP3 Player V2(網址：http://www.seeedstudio.com/depot/Grove-MP3-v20-p-2597.html)，不過筆者只有 Grove - Serial MP3 Player V1 的版本，所幸的是 V1 與 V2 功能差異不大，所以對讀者使用上，沒有太大差異。

Serial MP3 Player 模組

Grove - Serial MP3 Player 是 Seeed Studio 出廠的 Grove - Serial MP3 Player V1(網址：http://www.seeedstudio.com/depot/grove-serial-mp3-player-p-1542.html) (如下圖所示)，Grove-Serial MP3 Player 使用 WT5001 晶片為控制核心，產生高品質的音效，. 他支援 8KHZ~44.1kHZ 取樣頻率的 MP3 and WAV 音效檔，而控制方法只需兩條線。使用二個腳位串列埠來傳遞控制命令，控制音樂撥放。對於音效檔案，可以使用 Micro SD 記憶卡，所以說是一個非常方便的音效模組。

圖 19 Serial MP3 Player 模組

資料來源：Seeed Studio 官網：
http://www.seeedstudio.com/depot/grove-serial-mp3-player-p-1542.html

http://www.seeedstudio.com/wiki/Grove_-_Serial_MP3_Player

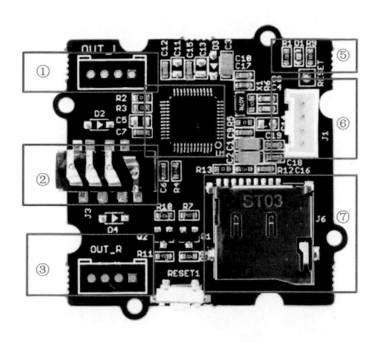

資料來源：Seeed Studio 官網：

http://www.seeedstudio.com/depot/grove-serial-mp3-player-p-1542.html

http://www.seeedstudio.com/wiki/Grove_-_Serial_MP3_Player

由上圖所示，下列為 Serial MP3 Player 的腳位資料：

- ◆ 1：左聲道腳位
- ◆ 2： 3.5mm 耳機插座
- ◆ 3：右聲道腳位
- ◆ 4： WT5001 晶片
- ◆ 5：LED Indicator：撥放燈指示，當燈亮就是音樂撥放
- ◆ 6：UART 控制腳位
- ◆ 2：SD Card: micro SD 記憶卡(小於 2GB 容量卡皆可)

電路組立

首先，組立 Serial MP3 Player 非常簡單，只要把 Serial MP3 Player 插入接轉插杜邦線(如上上圖所示)，如上圖所示，接入上圖.(6)的插槽，再將杜邦線接在 Arduino UNO 開發板電源(+5V、GND)與腳二、腳三(UART 用)，如下圖所示，把電源與 UART 腳位接好，就完成電路組立。

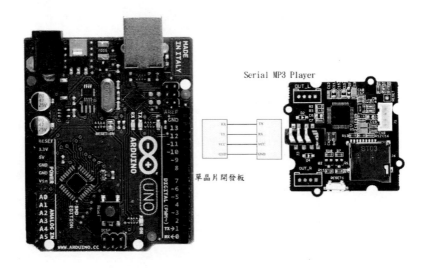

圖 20 Serial MP3 V1 接腳圖

請到筆者 Github：https://github.com/brucetsao/eSound/tree/master/MP3Files 或讀者有自己的音效檔，下載測試音效檔：0001.mp3 & 0002.mp3，並將該檔案存入 Micro SD 記憶卡的根目錄之下。

如下圖所示，為 Serial MP3 V1 Micro SD 卡插槽處，將安裝好音效檔的 Micro SD 記憶卡插入下圖紅框處，該模組只可以支援到 2G 容量的 Micro SD 記憶卡。

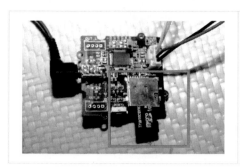

圖 21 Arduino Wave Shield SD 卡插槽處

我們請讀者鍵入Ｓｋｅｔｃｈ　ＩＤＥ軟體（軟體下載請到：https://www.arduino.cc/en/Main/Software），編譯完成後上傳到開發版進行測試。

表 11 Arduino Wave Shield 測試程式一

Arduino Wave Shield 測試程式一(Serial_MP3_Player)
| ```
/***********************************************************************
**/
//      Function: control the seeedstudio Grove MP3 player
//          Hardware: Grove - Serial MP3 Player
/***********************************************************************
****/
#include <SoftwareSerial.h>
SoftwareSerial mp3(2, 3);//modify this with the connector you are using.
void setup()
{
    mp3.begin(9600);
    Serial.begin(9600);
    delay(100);
    if (true ==SetPlayMode(0x01))
    Serial.println("Set The Play Mode to 0x01, Single Loop Mode.");
    else
    Serial.println("Playmode Set Error");
    PauseOnOffCurrentMusic();

}
void loop()
``` |
```

```
{
    SetPlayMode(0x01);
    delay(1000);
    SetMusicPlay(00,01);
    delay(1000);
    SetVolume(0x0E);
    while(1);
}
//Set the music index to play, the index is decided by the input sequence
//of the music;
//hbyte: the high byte of the index;
//lbyte: the low byte of the index;
boolean SetMusicPlay(uint8_t hbyte,uint8_t lbyte)
{
    mp3.write(0x7E);
    mp3.write(0x04);
    mp3.write(0xA0);
    mp3.write(hbyte);
    mp3.write(lbyte);
    mp3.write(0x7E);
    delay(10);
    while(mp3.available())
    {
        if (0xA0==mp3.read())
        return true;
        else
        return false;
    }
}
// Pause on/off   the current music
boolean PauseOnOffCurrentMusic(void)
{
    mp3.write(0x7E);
    mp3.write(0x02);
    mp3.write(0xA3);
    mp3.write(0x7E);
    delay(10);
    while(mp3.available())
    {
```

```
            if (0xA3==mp3.read())
            return true;
            else
            return false;
        }
}

//Set the volume, the range is 0x00 to 0x1F
boolean SetVolume(uint8_t volume)
{
    mp3.write(0x7E);
    mp3.write(0x03);
    mp3.write(0xA7);
    mp3.write(volume);
    mp3.write(0x7E);
    delay(10);
    while(mp3.available())
    {
        if (0xA7==mp3.read())
        return true;
        else
        return false;
    }
}

boolean SetPlayMode(uint8_t playmode)
{
    if
(((playmode==0x00)|(playmode==0x01)|(playmode==0x02)|(playmode==0x03))==false)
    {
        Serial.println("PlayMode Parameter Error! ");
        return false;
    }
    mp3.write(0x7E);
    mp3.write(0x03);
    mp3.write(0xA9);
    mp3.write(playmode);
    mp3.write(0x7E);
    delay(10);
```

```
    while(mp3.available())
    {
        if (0xA9==mp3.read())
        return true;
        else
        return false;
    }
    }
```

再將程式上傳到開發板之後，我們可以聽到插入耳機插孔的喇叭，可以發出對

應的音樂或音效。

函數用法

由下表所示，我們可以看到，控制 Serial_MP3_Player，只要用表中的命令，透

過串列埠傳送資料給 Serial_MP3_Player 的串列埠，就可以控制其音效，其他細節，

可以請讀者詳讀附錄相關資料。

表 12 Serial MP3 撥放命令表

Command Name	Command Format	Description
Pause	7E 02 A3 7E	您傳送這個命令，會使 Serial Player 暫停，再送一次命令，會繼續撥放 回傳：0xA3 為成功，其他為否
Stop	7E 02 A4 7E	您傳送這個命令，會使 Serial Player 停止撥放
Next	7E 02 A5 7E	您傳送這個命令，會使 Serial Player 撥放下一條歌

Previous	7E 02 A6 7E	您傳送這個命令，會使 Serial Player 撥放上一條歌
Volume control	7E 03 A7 1F 7E 7E 03 A7 *XX* 7E	設定播放放聲音大小，*XX* 表聲音大小，00 為無聲音，31 為最大聲音，必須用十六進位傳入，0x00 為無聲音，0x1F 為最大聲音 回傳：0xA7 為成功，其他為否
Assigned play mode	7E 03 A9 XX 7E	播放模式設定 XX=00 單純播放方式 回傳：0xA9 為成功，其他為否
		播放模式設定 XX=01 單首歌重複播放方式 回傳：0xA9 為成功，其他為否
		播放模式設定 XX=02 整片 CD/目錄重複播放方式. 回傳：0xA9 為成功，其他為否
		播放模式設定 XX=03 隨機播放方式 回傳：0xA9 為成功，其他為否

由於控制串列埠對有些讀者較為困難，本章的測試程式已將上表命令包成函數，讀者可以直接使用測試程式或使用測試程式在修改之。

為了更能了解 Serial MP3 的用法，本節介紹測試程式中該函式主要的用法：

設定 MP3 Player 模式

SetPlayMode(uint8_t playmode)

播放模式設定
- 00 單純播放方式
- 01 單首歌重複播放方式
- 02 整片 CD/目錄重複播放方式.
- 03 隨機播放方式

設定 MP3 Player 聲音大小

SetVolume(uint8_t volume)

設定播放放聲音大小，**XX** 表聲音大小，00 為無聲音，31 為最大聲音，必須用十六進位傳入，0x00 為無聲音，0x1F 為最大聲音

設定 MP3 Player 模式

SetMusicPlay(uint8_t hbyte,uint8_t lbyte)

設定 Serial Player 撥放哪一條歌，傳入兩個 Bytes，指定哪撥放哪一條歌。

uint8_t hbyte: 高位元組的資料;

uint8_t lbyte: 低位元組的資料

歌曲數= uint8_t hbyte: 高位元組的資料 * 256+ uint8_t hbyte: 高位元組的資料;

uint8_t lbyte: 低位元組的資料

設定 MP3 Player 暫停/繼續撥放

PauseOnOffCurrentMusic(void)

這個命令，會使 Serial Player 暫停，再送一次命令，會繼續撥放

Now you can hear songs stored in your SD card. And in the playing mode, the D1 indicator is on. If in the pause mode, the indicator will blink. More experience is waiting for you！

章節小結

本章節介紹 Seeed Studio 出廠的 Serial_MP3_Player V1，主要是讓讀者了解 Arduino 開發板如何透過 Seeed Studio 出廠的 Serial_MP3_Player V1，讓 Arduino 開發

板可以發出音效，透過以上章節的內容，一定可以一步一步的將音效輸出給予實作出來。

6

CHAPTER

DFPlayer Mini

DFRobot(網址：http://www.dfrobot.com/)出廠的 DFPlayer Mini 音效模組：
http://www.dfrobot.com/index.php?route=product/product&product_id=1121&search=
DFPlayer+Mini&description=true#.V6NtnPl97IU)，讀者可以從台灣 ICShop 的賣場：
http://www.icshop.com.tw/product_info.php/products_id/18039、德源科技的賣場：
http://goods.ruten.com.tw/item/show?21307231205130、鈺瀚網舖的賣場：
http://goods.ruten.com.tw/item/show?21627052903197、傑森創工坊的賣場：
http://goods.ruten.com.tw/item/show?21603717689257、都会明武电子的賣場：
https://world.taobao.com/item/520474174361.htm?fromSite=main&spm=a312a.770084
6.0.0.AnJFl8&_u=5vlvti95e7e，買到這篇音效擴充板，價格約台幣一、二百元上下，
可以說是價廉物美，CP 值超高的一塊音效模組。

DFPlayer Mini 模組

DFRobot(網址：http://www.dfrobot.com/)出廠的 DFPlayer Mini 音效模組：
http://www.dfrobot.com/index.php?route=product/product&product_id=1121&search=
DFPlayer+Mini&description=true#.V6NtnPl97IU)，(如下圖所示)，他支援 14KHz/16bit
WAV 音效檔與 32KHz/16bit AD4 音效檔，本張音效模組支援 DAC 輸出與 PWM
輸出，對於音效檔案，可以使用 SD/MMC 記憶卡，本模組只需要三個腳位就可以
控制音樂撥放，所以說是一個非常方便的音效模組。

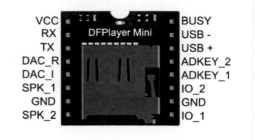

圖 22 DFPlayer Mini 音效模組

資料來源：DFRobot 官網：http://www.dfrobot.com/wiki/index.php/DFPlayer_Mini_SKU:DFR0299

下列為 DFPlayer Mini 模組的規格資料：

- 支援 FAT16 , FAT32 的 TF 卡擴充音效檔(32 GB 可用)。

- 支援下列 8/11.025/12/16/22.05/24/32/44.1/48 (kHz)取樣頻率的音效

- 支援 24 -bit DAC 音效輸出、動態可達 90dB , SNR 支援到 85dB。

- 多種控制模式可以選用：I/O 控制模式, 串列埠控制模式,按鈕控制模式

- 可用資料夾儲存音效檔，支援到 100 個資料夾，每一個資料夾可支援到 255 音效檔

- 支援 30 階層的音量調整功能、6 階層的 EQ 調整功能

- 大小: 20mm*20mm

- 重量: 20g

電路組立

首先，組立 DFPlayer Mini 模組非常簡單，只要把 DFPlayer Mini 模組的 VCC/GND 插入接轉插杜邦線(如下圖所示)，接在 Arduino UNO 開發板電源(+5V、GND)，再將 DFPlayer Mini 模組 RX/TX 用杜邦線接在 Arduino UNO 開發板腳二、腳三(UART 用)，如下圖所示，把電源與 UART 腳位接好，就完成電路組立。

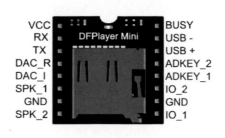

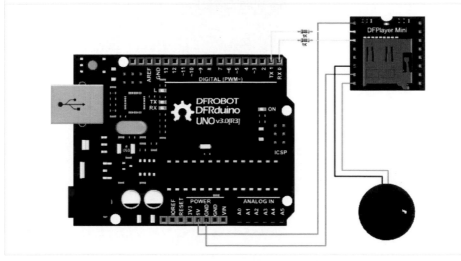

圖 23 DFPlayer Mini 音效模組接腳圖

資料來源：http://www.dfrobot.com/wiki/index.php/DFPlayer_Mini_SKU:DFR0299

請到筆者 Github：https://github.com/brucetsao/eSound/tree/master/MP3Files 或讀者有自己的音效檔，下載測試音效檔：0001.mp3 & 0002.mp3，並將該檔案存入 TF 記憶卡的根目錄下的 mp3 資料夾中。

如下圖所示，為 DFPlayer Mini 模組 TF 記憶卡插槽處，將安裝好音效檔的 TF 記憶卡插入下圖紅框處，該模組只可以支援到 32G 容量的 TF 記憶卡。

如下圖所示，DFPlayer Mini 模組 TF 記憶卡插槽處，將安裝好音效檔的 TF 記憶卡插入下圖紅框處，該模組只可以支援到 32G 容量的 TF 記憶卡。

圖 24 DFPlayer Mini 模組 TF 卡插槽處

我們請讀者鍵入 Ｓｋｅｔｃｈ ＩＤＥ軟體（軟體下載請到：https://www.arduino.cc/en/Main/Software)，編譯完成後上傳到開發版進行測試。

表 13 DFPlayer Mini 測試程式一

DFPlayer Mini 測試程式一(DFPlayer01)
/*   *   Copyright: DFRobot

```
 *    name:         DFPlayer_Mini_Mp3 sample code
 *    Author:       lisper <lisper.li@dfrobot.com>
 *    Date:         2014-05-30
 *    Description:   sample code for DFPlayer Mini, this code is test on Uno
 *                  note: mp3 file must put into mp3 folder in your tf card
 */


#include <SoftwareSerial.h>
#include <DFPlayer_Mini_Mp3.h>

SoftwareSerial mySerial(2, 3); // RX, TX
void setup () {
    Serial.begin (9600);
    mySerial.begin (9600);
    mp3_set_serial (mySerial);       //set Serial for DFPlayer-mini mp3
module
    delay(1);                        // delay 1ms to set volume
    mp3_set_volume (30);             // value 0~30
}

void loop () {
    mp3_play (1);
    while(1) ;

}

/*
    mp3_play ();        //start play
    mp3_play (5);       //play "mp3/0005.mp3"
    mp3_next ();        //play next
    mp3_prev ();        //play previous
    mp3_set_volume (uint16_t volume);       //0~30
    mp3_set_EQ ();     //0~5
    mp3_pause ();
    mp3_stop ();
    void mp3_get_state ();      //send get state command
    void mp3_get_volume ();
    void mp3_get_u_sum ();
```

```
    void mp3_get_tf_sum ();
    void mp3_get_flash_sum ();
    void mp3_get_tf_current ();
    void mp3_get_u_current ();
    void mp3_get_flash_current ();
    void mp3_single_loop (boolean state);      //set single loop
    void mp3_DAC (boolean state);
    void mp3_random_play ();
*/
```

參考網址：http://www.dfrobot.com/wiki/index.php/DFPlayer_Mini_SKU:DFR0299

程式下載網址：https://github.com/brucetsao/eSound/

　　再將程式上傳到開發板之後，我們可以聽到插入耳機插孔的喇叭，可以發出對應的音樂或音效。

播放模式介紹

　　對於播放更進階的播放方式，讀者可以參考下表，適用下列串列埠命令來播放更進階的功能。

表 14 播放指令介紹

命令	命令介紹	參數(16 位元)
0x01	下一首	
0x02	上一首	
0x03	播放指定曲(NUM)	0-2999
0x04	增加音量	
0x05	減少音量	
0x06	指定音量	0-30
0x07	指定 EQ 0/1/2/3/4/5	Normal/Pop/Rock/Jazz/Classic/Bass
0x08	設定播放模式(0/1/2/3)	repeat/folder repeat/single repeat/random
0x09	設定播放資料來源(0/1/2/3/4)	U/TF/AUX/SLEEP/FLASH
0x0A	進入休眠	

0x0B	恢復正常工作(從休眠狀態)	
0x0C	重置模組	
0x0D	播放	
0x0E	暫停	
0x0F	指定播放資料夾	1-10（need to set by user）
0x10	調整音量	[DH=1:Open volume adjust][DL:set volume gain 0-31]
0x11	設定重複模式	[1:start repeat play][0:stop play]
0x12	指定播放 MP3 資料夾	0-9999
0x13	Commercials	0-9999
0x14	Support 15 folder	See detailed description below
0x15	停止回放,背景播放	
0x16	停止回放	

參考網址：http://www.dfrobot.com/wiki/index.php/DFPlayer_Mini_SKU:DFR0299

讀者可以參考下表，了解 DFPlayer Mini 模組目前狀態，可以透過下列串列埠
命令來取得模組目前狀態。

表 15 查詢令命

命令	命令介紹	參數(16 位元)
0x3C	保留	
0x3D	保留	
0x3E	保留	
0x3F	傳送初始化參數	0-0x0F（each bit represent one device of the low-four bits）
0x40	回應錯誤，需要傳輸資料	
0x41	回應	
0x42	查詢目前狀態	
0x43	查詢目前音量	
0x44	查詢目前 EQ 狀態	
0x45	查詢目前播放模式	This version retains this feature
0x46	查詢目前韌體版本	This version retains this feature

0x47	查詢目前 TF 記憶卡音效檔總數	
0x48	查詢目前 U-disk 音效檔總數	
0x49	查詢目前 FLASH 音效檔總數	
0x4A	繼續	
0x4B	查詢目前播放 TF 記憶卡音效檔的檔案號碼	
0x4C	查詢目前播放 U-disk 音效檔的檔案號碼	
0x4D	查詢目前播放 FLASH 音效檔的檔案號碼	

參考網址：http://www.dfrobot.com/wiki/index.php/DFPlayer_Mini_SKU:DFR0299

組合鍵播放模式

　　DFPlayer Mini 模組可以不需要透過軟體驅動，可以直接使用硬體線路，將下圖所示之按鍵與對應電路，將他接在 DFPlayer Mini 模組之下，就可以將 DFPlayer Mini 模組轉換成一個很棒的 MP3 播放器。

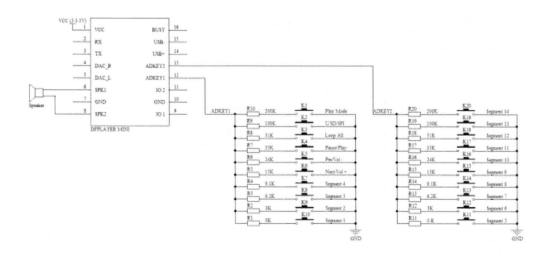

圖 25 DFPlayer Mini 模組硬體按鍵模式電路圖範例

參考網址：http://www.dfrobot.com/wiki/index.php/DFPlayer_Mini_SKU:DFR0299

I/O Mode

　　DFPlayer Mini 模組可以透過軟體驅動,直接驅動播放音樂的所有功能,對於簡單的聲音、上下曲等功能,只要加上下圖所示之按鍵電路,就可以整合成軟硬體綜合控制的方式,也可以不加入下圖所示之按鍵電路,純用軟體控制,這是一個可以用多種方式控制的模組。

- Refer diagram

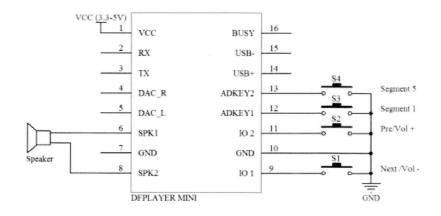

圖 26 DFPlayer Mini 模組軟體附加控制鍵模式電路圖範例

參考網址:http://www.dfrobot.com/wiki/index.php/DFPlayer_Mini_SKU:DFR0299

章節小結

　　本章節介紹 DFRobot 出廠的 DFPlayer Mini 音效模組,主要是讓讀者了解 Arduino 開發板如何透過 DFPlayer Mini 讓 Arduino 開發板可以發出音效,透過以上章節的內容,一定可以一步一步的將音效輸出給予實作出來。

本書總結

　　筆者對於 Arduino 相關的書籍，也出版許多書籍，感謝許多有心的讀者提供筆者許多寶貴的意見與建議，作者群不勝感激，許多讀者希望筆者可以推出更多的入門書籍給更多想要進入『Arduino』、『Maker』這個未來大趨勢，所有才有這個入門系列的產生。

　　本系列叢書的特色是一步一步教導大家使用更基礎的東西，來累積各位的基礎能力，讓大家能更在 Maker 自造者運動中，可以拔的頭籌，所以本系列是一個永不結束的系列，只要更多的東西被製造出來，相信筆者會更衷心的希望與各位永遠在這條 Maker 路上與大家同行。

作者介紹

曹永忠 (Yung-Chung Tsao) ，目前為自由作家暨專業 Maker，專研於軟體工程、軟體開發與設計、物件導向程式設計，商品攝影及人像攝影。長期投入創客運動、資訊系統設計與開發、企業應用系統開發、軟體工程、新產品開發管理、商品及人像攝影等領域，並持續發表作品及相關專業著作。

Email:prgbruce@gmail.com

Line ID：dr.brucetsao

部落格：http://taiwanarduino.blogspot.tw/

書本範例網址：https://github.com/brucetsao/eSound

臉書社群(Arduino.Taiwan)：

https://www.facebook.com/groups/Arduino.Taiwan/

Youtube：https://www.youtube.com/channel/UCcYG2yY_u0m1aotcA4hrRgQ

許智誠 (Chih-Cheng Hsu)，美國加州大學洛杉磯分校(UCLA) 資訊工程系博士，曾任職於美國 IBM 等軟體公司多年，現任教於中央大學資訊管理學系專任副教授，主要研究為軟體工程、設計流程與自動化、數位教學、雲端裝置、多層式網頁系統、系統整合。

Email: khsu@mgt.ncu.edu.tw

蔡英德 (Yin-Te Tsai)，國立清華大學資訊科學系博士，目前是靜宜大學資訊傳播工程學系教授、靜宜大學計算機及通訊中心主任，主要研究為演算法設計與分析、生物資訊、軟體開發、視障輔具設計與開發。

Email:yttsai@pu.edu.tw

附錄

WT588D-U 語音模組(英文版)

The Instructions of WT588D-U Voice Module

www.elechouse.com

1、 Product features

➤ Package of 28 pins module, which can be replaced memory for gainting different lengths of storage time.

➤ Support SPI-Flash, which 's capacity is 2M bit ~ 32M (Note: 1byte = 8bit)

➤ WT588D-20SS voice used as a control core chip.

➤ Embedded human voice processor, feel very natural and sweet.

➤ Good audio quality output for 13Bit/DA converter and 12Bit/PWM processing of audio.

➤ Support for loading 6K ~ 22KHz audio sampling rate.

➤ PWM output can directly promote 0.5W/8Ω speakers and plenty of current.

➤ Support DAC / PWM output

➤ Support for loading WAV audio format.

➤ Support key control mode, one-wire serial control mode, three-wire serial control mode.

➤ A variety of IO trigger ways can be seted to in button control mode.

➤ The way of BUSY signal output can be set in a random manner.

➤ Loading no more than 500 segments voice for editing.

➤ Address bit is controled by 220 segments voice, but a single address bit just can load up to 128.

➤ Voice player to enter the sleep mode to stop immediately.

➤ It is simple interface and convenient because of using WT588D Voice Chip that benefited to exert its functions.

➤ A lot of operations can be finished in software. Such as setup control mode, inserting voice, compositing voice, calling voice, etc.

➤ Free to insert mute, mute time range 10ms ~ 25min.

➤ Support online USB download / offline USB download. What's more, it also can download data to SPI-Flash even if WT588D-U are working.

➤ Operating voltage: DC2.8V ~ 5.5V.

➢ dormant current less than 10uA

➢ Powerful anti-jamming. Widely used in the industrial field.

2、 Functional Description

Button control mode is flexible to trigger and free to set any button to re-trigger .There are 15 trigger ways. Including trigger Impulse Repetition , trigger pluse Without Repetition , invalidation keys, no cycle Level , Recycled Level , Non-Maintained Cycle Level, Non-Cycle for The Last One ,Non-Cycle for The Next One , Cycle for The Last One , Cycle for The Next One, Volume +, Volume -, play / pause, stop, play / stop, etc . One-wire serial control mode and three-wire serial control mode, not only can control voice play, stop, loop play and volume size by the MCU, but also can direct triggering any voice in address bit from 0 to 219.

3、 application scopes

Widely range of applications. Almost related to all the voice places, such as Stop devices, annunciators , reminder, alarm clock, learning machine, intelligent home appliances, therapeutic equipment, electronic toys, telecommunications, reversing radar and a variety of automatic control devices, etc. Technology meet up to the requirements of industries application.

4、 Application block diagram

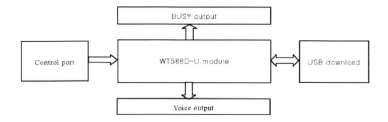

5

```
 1  NC        VDD-USB  28
 2  NC        D+       27
 3  NC        D-       26
 4  NC        GND      25
 5  NC        NC       24
 6  NC        NC       23
 7  RESET     VDD      22
 8  DAC       BUSY     21
 9  PWM +     VCC      20
10  PWM -     P00      19
11  P14       P01      18
12  P13       P02      17
13  P16       P03      16
14  GND       P15      15
```

Pin Description

Package pins	Pins mark	Bnet	Functional Description
1	NC	NC	blank
2	NC	NC	blank
3	NC	NC	blank
4	NC	NC	blank
5	NC	NC	blank
6	NC	NC	blank
7	RESET	RESET	Reset pin
8	DAC	DAC	DAC Audio output pin, need an external amplifier to drive speaker
9	PWM+	PWM+	PWM+ Audio output pin, which can directly drive speaker with the PWM−
10	PWM−	PWM−	PWM− Audio output pin, which can directly drive speaker with the PWM+
11	P14	SPI−FLASH_DI	use for external download manager
12	P13	SPI−FLASH_DO	use for external download manager
13	P16	SPI−FLASH_CLK	use for external download manager
14	GND	GND	GND
15	P15	SPI−FLASH_CS	use for external download manager
16	P03	K4/CLK/DATA	Button / three-wire clock / one-wire data input pin
17	P02	K3/CS	Button / three-wire chip input pin
18	P01	K2/DATA	Button / three-wire data input pin
19	P00	K1	Button
20	VCC	VCC	Analog power supply input pin

21	BUSY	BUSY	busy signal output pin
22	VDD	VDD	Data power supply input pin
23	NC	NC	blank
24	NC	NC	blank
25	GND	GND	USB GND
26	D-	USB_DATA-	USB data
27	D+	USB_DATA+	USB data
28	VDD_USB	VDD_USB	USB power positive

Note:Pin25,26,27,28 use to download for other USB slot.

6、electrical parameters

$V_{DD} - V_{SS} = 4.5V$, TA $= 25°C$，No load

Parameter	marker	Environmental conditions	min	Typical	Max	Units
Operating voltage	V_{DD}	F_{sys}=8MHz	2.8		5.5	V
Operating Current	I_{OP1}	No load	–	4.5	5.5	mA
Stop current	I_{DD2}	No load	–	1	2	uA
Dormancy current-mode	I_{OP2}	No load	–	650		uA
Low-voltage input	V_{IL}	All pin input	V_{SS}	–	$0.3V_{DD}$	V
high-voltage input	V_{IH}	All pin input	$0.7V_{DD}$	–	V_{DD}	V
input currentBP1, BP2，RESET	I_{IN1}	V_{IN}=0V Pull-up resistance=500KΩ	–5	–9	–14	uA
input currentBP1, BP2，RESET	I_{IN2}	V_{IN}=0V Pull-up resistance=150KΩ	–15	–30	–45	uA
output current (BP0)	I_{OL}	VDD=3V，VOUT=0.4V	8	12	–	mA
	I_{OH}	VDD=3V，VOUT=2.6V	–4	–6	–	mA
	I_{OL}	VDD=4.5V，VOUT=1.0V	–	25	–	mA
	I_{OH}	VDD=4.5V，VOUT=2.6V	–	12	–	mA
output current (BP1)	I_{OL}	VDD=3V，VOUT=0.4V	4	10	–	mA
	I_{OH}	VDD=3V，VOUT=2.6V	–4	–6	–	mA
output current PWM+/PWM–	I_{OL1}	RL=8Ω	+200	–	–	mA
	I_{OH1}	【PWM+】 ── 【RL】 ── 【PWM–】	–200	–	–	mA
DAC Max current	I_{DAC}	RL=100Ω	–2.4	–3.0	–3.6	mA
			–4.0	–5.0	–6.0	

Pull-up resistor test	R_{PL}		75	150	225	

7、Absolute limits of the environment parameters

parameters	marker	Environmental conditions	rating	units
power	$V_{DD} - V_{SS}$	–	$-0.3 \sim +7.0$	V
Input voltage	V_{IN}	input	$V_{SS}-0.3 \sim V_{DD}+0.3$	V
Storage temperature	TSTG	–	$-55 \sim +150$	°C
Used temperature	T_{OPR}	–	$-40 \sim +85$	°C

8、control mode

8.1、Buttons Control Mode

Pins can directly trigger a function of chip to work. Each pin of the trigger can be set individually. Shockproof time of this mode time is about 10ms. There are 15 trigger ways. Including trigger Impulse Repetition , trigger pluse Without Repetition , invalidation keys, no cycle Level , Recycled Level , Non-Maintained Cycle Level, Non-Cycle for The Last Tone ,Non-Cycle for The Next Tone , Cycle for The Last Tone , Cycle for The Next Tone, Volume +, Volume -, play / pause, stop, play / stop, etc. see the following trigger timing diagram. For more details, see the following chart:

8.1.1、Trigger Impulse Repetition

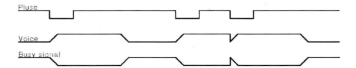

Note: Negative trigger pulse. When the I / O port inspects the falling edge (for example, the I / O port click short-circuit to GND), Voice will be broadcast .If do that again when the Voice are still playing, the

-第 7 页-

voice will be interrupted and replay. Therefore, it will be replay as long as has falling edge signal.

8.1.2、 Trigger Impulse Without Repetition

Note: Negative trigger pulse. When the I / O port inspects the falling edge (for example, the I / O port click short-circuit to GND), Voice will be broadcast. If do that again when the voice are still playing , the voice will not be interrupted and continue to broadcast. To be valid unless the voice at an end and inspects the falling edge.

8.1.3、 Recycled Level

Note: High level stops when the I / O port is low and keep play. Continue to keep a low level even if the first time is over. It will go along replay until change into high level. Low level has sound. High level hasn't.

8.1.4、 No Cycle Level

Note: Trigger level. High level stops when the I / O port is low and keep play. I will be not Continue to play even if the first time is over and keep a low level. The voice just play one time after being triggered.

If you need to replay, please make the I / O port at high level, and then pull low. Finally, keep it at low level .The end.

8.1.5、 Non-Maintained Cycle Level

Note: Negative Pulse /trigger Level. When the I / O port at low level and keep playing, at the same time, high level don't stop until the voice is over. When the end of the first time. If keep at the low level, it will continue to repeat .If not, when finish it will stop automatically.

8.1.6、 Play/Stop

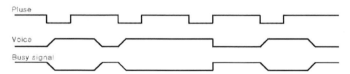

Note: Negative trigger pulse. Negative pulse starts to play when the next one stop. Whether the voice is in play or not must in accordance with this regulation.

8.1.7、 Non-Cycle for The Next Tone

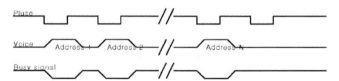

Note: Negative trigger pulse. Trigger with a button to play a sound. A pulse plays a piece, the next pulse plays the next piece .It doesn't stop until the last piece is finished. Repeat the same operation. Can only play to the last.

8.1.8、 Non-Cycle for The Last Tone

Note: Negative trigger pulse. Trigger with a button to play a sound. A pulse plays a piece, the next pulse plays the last piece. No longer trigger forward when the front voice is over. Repeat the operation, can only play to the last.

8.1.9、 Cycle for The Next Tone

Note: Negative trigger pulse. Trigger with a button to play a sound. A pulse plays a piece , the next pulse plays the next piece. Repeat the operation. It will start again from the first piece when the last shows off. Loop continuously.

8.1.10、 Cycle for The Last Tone

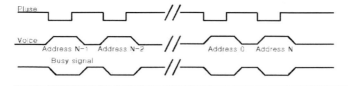

Note: Negative trigger pulse. Trigger with a button to play a sound. A pulse plays a piece, the next pulse plays the last piece. Repeat the operation. It will start again from the last piece when the front shows off. Loop continuously.

8.1.11、 Pause

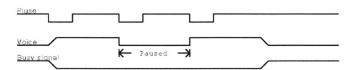

Note: Negative trigger pulse. The first pulse voice is playing but in a suspended state. The second pulse still working, which triggers the suspension of the voice. BUSY remain in this state.

8.1.12、 Stop

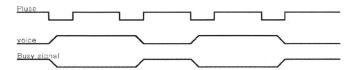

Note: Negative trigger pulse. Stopped the voice, which is playing . Trigger once again invalidly when the voice is stopped.

8.2、 One-Wire Serial Port Control Mode

Send data through a data line. One-wire serial port can control voice play, stop, volume adjustment and directly trigger, etc. P00 ~ P02 I/O port can select screen or any trigger mode.

8.2.1、 Port Allocation Table

I/O 口	P00	P01	P02	P03

function	Key-press K1	Key-pressK 2	Key-pressK3	DATA

8.2.2、 Order and Speech Cording

Command code	Functions	Descriptions
E0H~E7H	volume adjustment	8 volume can be adjusted, E0H is minimum, E7H is the largest volume when working or standby.
F2H	Cycle play	the current voice addresses can be recycled When working.
FEH	Stop playing	Voice command to stop playing

8.2.3、 Voice Address Correspondence

Data (hex)	functions
00H	Play the zero piece voice
01H	Play the first piece voice
02H	Play the second piece voice
......	......
D9H	Play the 217th piece voice
DAH	Play the 218th piece voice
DBH	Play the 219th piece voice

8.2.4、 Control Time Sequence Chart

One-wire serial port only through a data communication line control time sequence. According to different duty cycle of levels represent different data bit. Firstly, data signals is drawned down 5ms, and then send data .The duty cycle of High level and low level 1:3 means data bit 0, if 3:1 means data bit 1, high in the former. Data signals send from low to high. When Send data, you just send address datum directly can trigger to play voice without sending command code and instruction. D0 ~ D7 means an address or command data. 00H ~ DBH of data send address order. E0H ~ E7H is volume adjustment order.F2H is Loop orders. FEH orders to stop playing. Details of time sequence in the following diagram:

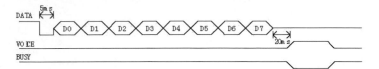

Description: WT588D−U can not enter dormant state under the one-wire serial interface. Therefore, using with caution when battery-powered .DATA is a communications line for one-wire serial interface, WT588D-U voice module begins to send data signals after current is switched on and wait 17ms.BUSY voice for the busy signal output. Wait for 20ms Data after sent successfully. And BUSY output will be to respond. Details of data bit duty cycle in the following chart:

High level：Low level=1：3 means 0   High level：Low level =3：1 means 1

8.2.5、 The Example of 1-Wire Serial Port Control Time Sequence

For example, Send time sequence of data 9CH chart in one-Wire Serial Port Control mode is show in figure:

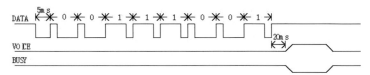

8.2.6、 Models of Procedure

Master SCM：PIC16F54, Clock frequency :4MHz

```
Send one-line (unsigned char addr)
 {
sda=0;
delay1ms(5);          /* Data signals at low   level 5ms */
for(i=0;i<8;i++)
 {   sda=1;
if(addr & 1)
```

```
{ delay1us(600);        /* High level: Low level =600us: 200us, means data=1 */
sda=0;
delay1us(200); }
else {
delay1us(600);         /* High level: Low level =200us: 600us, means data=0 */
sda=0;
delay1us(200); }
addr>>=1; }
sda=1; }
```

8.3、Three-Wire Serial Control mode

CS, DATA and CLK are composed of Three-Wire Serial Control mode .Time sequence according to SPI communication. Three-wire serial port can control command control and voice broadcast. All key-presses are not valid in the three-wire serial mode.

8.3.1、Port Allocation Methods

I/O 口	P00	P01	P02	P03
Functions	---	DATA	CS	CLK

8.3.2、Voice and Command Code Corresponding to Table

Command Code	Functions	Description
E0H ~ E7H	Volume adjustment	8 volume can be adjusted, E0H is minimum, E7H is the largest volume when working or standby.
F2H	Cycle play	The current voice addresses can be recycled When working.
FEH	Stop playing	Voice command to stop playing

8.3.3、Voice Address Corresponds

data（hex）	functions
00H	Play the zero piece voice
01H	Play the first piece voice
02H	Play the second piece voice

-第 14 页-

-95-

... ...	... ...
D9H	Play the 217th piece voice
DAH	Play the 218th piece voice
DBH	Play the 219th piece voice

8.3.4、 Three-Wire Serial Port Control Time Sequence

CS, CLK and DATA pins are composed of Three-Wire Serial Control mode .Time sequence follows to SPI communication. CS downs to 5ms in order to wake-up WT588D-U voice module. Low bit receives data at the rising edge of CLK in the first place. Clock cycles between the range of 100us ~ 2ms, recommended 300us. The BUSY voice outputs in response to the successful reception of data. Data signals send from low to high. When Send data, you just send address datum directly can trigger to play voice without sending command code and instruction. D0 ~ D7 means an address or command data. 00H ~ DBH of data send address order. E0H ~ E7H is volume adjustment order.F2H is Loop orders. FEH orders to stop play. Details of time sequence in the following diagram:

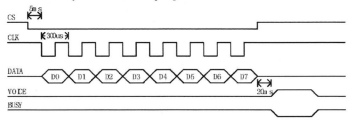

Description: WT588D-U voice module begins to send data signals after current is switched on and wairt 17ms.

8.3.5、 Models of Procedure

（Master SCM PIC16F54， System frequency 4MHz）
Send threelines(unsigned char addr)
{ cs=0;
delay1ms(5);   /* Chip select signal keep low level 2ms */
    for(i=0;i<8;i++)
    {   scl=0;
    if(addr & 1)sda=1;
    else sda=0;
    addr>>=1;

```
delaylus(300);          /* Clock cycle 300us */
scl=1;
delaylus(300);   }
cs=1;}
```

9、 Typical Application Circuit

9.1、 Typical Application Circuit of Key to Control(PWM output, 5V Supply)

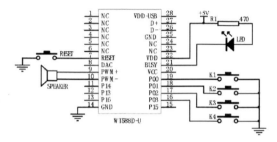

9.2、 Typical Application Circuit of Keys to Control (PWM output, 3.3V Supply)

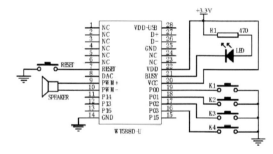

9.3、 Typical Application Circuit of Keys to Control（DAC output）

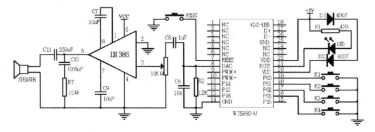

Note: DAC output port together with the ground, which connect with a 1.2K resistor and capacitor 104. when use DAC output way, and then the audio signal re-entering amplifier part, as circuit diagram of R2, R6 shown.

9.4、 Typical Application of one-line Serial Control Circuit (PWM out)

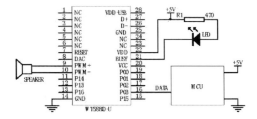

9.5、 Typical Application of First-line Serial Control Circuit（DAC Output）

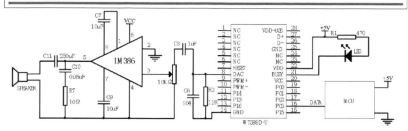

Note: DAC output port together with the land , which connect with a 1.2K resistor and capacitor 104. when use DAC output way, and then the audio signal re-entering amplifier part, as circuit diagram of R2, R6 shown.

9.6、0ne-Wire Serial Port MCU5V Power/ Module 3.3V Power Supply Application circuit (PWM output)

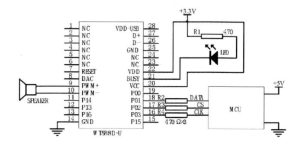

9.7、 Three-Wire Serial Control of Typical Application circuit (PWM output)

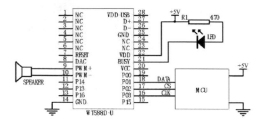

9.8、 Three-wire serial control of a typical application circuit (DAC output)

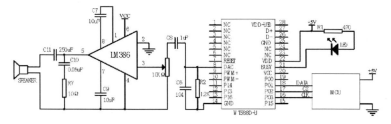

Note: DAC output port together with the ground, which connect with a 1.2K resistor and capacitor 104.

When use DAC output way, and then the audio signal re-entering amplifier part, as circuit diagram of R2, R6 shown.

9.9、 Three-wire serial MCU5V power / module 3.3V power supply application circuit (PWM output)

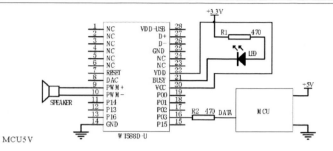

MCU5V

10、 Control procedures

10.1、 One-wire serial control of assembler

Description: This procedure is test program. Please change the IO port of MCU according to Practical application.

```
 ORG 0000H
        KEY EQU P1.1      ;   Button pin
        SDA EQU P3.0      ;   Data pin
DAIFAZHI EQU 50H          ;   A temporary address for Code value
        MOV DAIFAZHI,#0H;   Code made the initial value of 0
        MOV R5,#8         ;   8-bit Circulation of Fat Code

MAIN:
        JB KEY,MAIN
        MOV R6,#20          ;Delay 20MS
        LCALL DELAY1MS
        JB KEY,MAIN         ; Buffeting button to judgment
        JNB KEY,$           ;Wait for button release
        LCALL one-line      ; Transfer one-wire fat code Subroutine
        INC DAIFAZHI        ; Code value plus 1 fat
        MOV A,DAIFAZHI
        CJNE A,#220,XX2 ,Whether reach max 220 of the Voice paragraph or not
XX2: JC XX3
        MOV DAIFAZHI,#0H
XX3: LJMP MAIN
```

```
One-line:                ;//// one-wire fat code Subroutine
            CLR SDA
            MOV R6,#5        ; Delay 5MS
            LCALL DELAY1MS
                MOV A, DAIFAZHI
        LOOP:     SETB SDA
                RRC A
                JNC DIDIANPIN   ; High level pulse   High: Low=3:1
            LCALL DELAY200US
LCALL DELAY200US
            LCALL DELAY200US
                CLR   SDA
            LCALL DELAY200US
            LJMP LOOP1
    DIDIANPIN                 ; Low level pulse   High: Low =1:3
            LCALL DELAY200US
                CLR SDA
            LCALL DELAY200US
LCALL DELAY200US
            LCALL DELAY200US
        LOOP1:   DJNZ R5,LOOP
                MOV R5,#08H
            SETB SDA
                RET
DELAY200US:   MOV R6,#100    ; Delay Subroutine 400US
                DJNZ R6,$
        RET
    DELAY1MS:                 ; Delay Subroutine 1ms, help R6 evaluate, Modified to extend the
time

        L1:   MOV R7,#248
            DJNZ R7,$
            DJNZ R6,L1
            RET
END
```

10.2、One-wire serial control of C-voice procedures

Description: This procedure is test program.Please change the IO port of MCU according to Practical application.

```
#include <at89x2051.H>
sbit KEY=P1^1; /* The 2nd of P1 port is P1_1 */
```

```
sbit SDA=P3^0; /* The 4th of P3 port is P3_0 P3_0 */
void delay1ms(unsigned char count) //1MS delay time Subroutine
{
    unsigned char i,j,k;
    for(k=count;k>0;k--)
        for(i=2;i>0;i--)
        for(j=248;j>0;j--);
}

void delay100us(unsigned char count)   //100US Delay time Subroutine
{ unsigned char i;
    unsigned char j;
        for(i=count;i>0;i--)
        for(j=50;j>0;j--);
}

Send_oneline(unsigned char addr)
{
    unsigned char i;
    SDA=0;
    delay1ms(5);              /* delay 5ms */
    for(i=0;i<8;i++)
        {SDA=1;
        if(addr & 1)
            {delay100us(6);          /* 600us */
            SDA=0;
            delay100us(2);     /* 200us */
            }
        else {
            delay100us(2);      /* 200us */
            SDA=0;
            delay100us(6);     /* 600us */
            }
        addr>>=1; }
        SDA=1;
}

main()
{unsigned char FD=0;
    P3=0XFF;
    while(1)
        {
```

```
                if(KEY==0)
                    {
                    delay1ms(10);
                    if(KEY==0)        // Increase Code value of fat by button P1.1.
                        {
                        Send online (FD);
                    FD++;
                        if(FD==220) // One-wire Serial port, the voice segment up to a maximum of 220
        {
                            FD=0;
                        }
                        while(KEY==0);   // Waiting for button release in order to avoid Miscarriage of justice
                    }
            }
    }
```

10.3、Three-wire serial control of assembler

Description: This procedure is test program.Please change the IO port of MCU according to Practical application.

```
                ORG 0000H
            KEY EQU P1.1      ; Button pin
            CS   EQU P3.1     ;CS trigger pin
            SCL EQU P3.2      ;Clock pin
            SDA EQU P3.0      ;Data pin
DAIFAZHI EQU 50H;A temporary address for Code value
                MOV DAIFAZHI,#0H;Code made the initial value of 0
            MOV R5,#8          ;Code made the initial value of 0

MAIN:
            JB KEY,MAIN
            MOV R6,#20        ;Dalay time 20MS
            LCALL DELAY1MS
            JB KEY,MAIN          ; Buffeting button to judgment
            JNB KEY,$            ; Wait for button release
            LCALL THREE_LINE; Transfer three-wire fat code Subroutine
            INC DAIFAZHI        , Code value plus 1 fat
            MOV A,DAIFAZHI
            CJNE A,#220,XX2 ; Whether reach max 220 of the Voice paragraph or not
XX2: JC XX3
            MOV DAIFAZHI,#0H
```

XX3: LJMP MAIN

```
THREE_LINE:              ;//// three-wire fat code Subroutine
            CLR CS
            MOV R6,#5        ; Dalay time 5MS
            LCALL DELAY1MS
                MOV A,DAIFAZHI
        LOOP:
                CLR SCL
                RRC A
                MOV SDA,C
                LCALL DELAY50US
                SETB SCL
            LCALL DELAY50US
                DJNZ R5,LOOP
                MOV R5,#08H
                SETB CS
                RET
DELAY50US:      MOV R6,#150        ; Subroutine of dalay time 300US
                DJNZ R6,$
            RET
DELAY1MS:                          ; Delay Subroutine 1ms, help R6 evaluate. Modified to extend the time

        L1: MOV R7,#248
        L2: NOP
            NOP
            DJNZ R7,L2
            DJNZ R6,L1
            RET
                END
```

10.4、 Three-wire serial control of C-voice procedures

Description: This procedure is test program.Please change the IO port of MCU according to Practical application.
```
#include <at89x51.H>
sbit KEY=P1^1; /*      The 2nd of P1 port is P1_1 */
sbit   CS=P3^1; /*     The 3rd of P3 port is P3_1 */
sbit SCL=P3^2; /*      The 4th of P3 port is P3_2 */
sbit SDA=P3^0; /*      The 5th of P3 port isP3_0 */
```

```
//sbit DENG=P3^7; /*  The 6th of P3 port is P3_5- */
void delay1ms(unsigned char count) //1MS Dalay time subroutine
{
      unsigned char i,j,k;
      for(k=count;k>0;k--)
           for(i=2;i>0;i--)
           for(j=248;j>0;j--);
}


void delay100us(void)   //100US Dalay time subroutine
{
      unsigned char j;
           for(j=50;j>0;j--);
}


Send_threelines(unsigned char addr) // three-wire fat code Subroutine
      {unsigned char i;
           CS=0;
           delay1ms(5);
      for(i=0;i<8;i++)
           {SCL=0;
           if(addr & 1)SDA=1;
           else SDA=0;
           addr>>=1;
           Delay300us();   /* 300us */
           SCL=1;
           Delay300us();
           }
           CS=1;
      }


main()
{unsigned char FD=0;
           P3=0XFF;
           while(1)
           {
           if(KEY==0)
             {
                delay1ms(20);
                if(KEY==0)     //Increase Code value of fat by button P1.1.
                {
                   Send three-line (FD);
```

-第 25页-

```
          FD++;
              if(FD==220//Three-wire Serial port, the voice segment up to a maximum of 220
      {
              FD=0;
          }
          while(KEY==0);    // Waiting for button release in order to avoid Miscarriage of justice
              }
          }
      }
}
```

11、Package size Figure

Units： mm

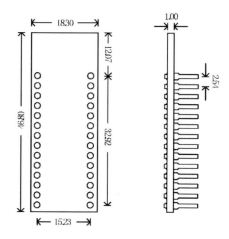

資料來源：

http://www.elechouse.com/elechouse/images/product/MP3%20WT588D%20USB%20modul
e/WT588D-U%20Voice%20Module.pdf

WT588D-語音模組(中文版)

WT588D 語音模組

使用說明書

目　　　錄

2

1.功能概述

WT588D 語音模組是一款功能強大的可重複擦除燒寫的語音單晶片機。WT588D 讓語音晶片不再為控制方式而尋找合適的外圍單片機電路，高度集成的單晶片技術足於取代複雜的週邊控制電路。

配套 WT588D VoiceChip 燒錄操作軟體可隨意更換 WT588D 語音模組的任何一種控制模式，把資訊下載到 SPI-Flash 上即可。軟體操作方式簡潔易懂，攝合了語音組合技術，大大減少了語音編輯的時間。

完全支持線上下載，即便是 WT588D 通電的情況下，一樣可以通過下載器給關聯的 SPI-Flash 下載資訊，給 WT588D 語音晶片電路重定一下，就能更新到剛下載進來的控制模式。支援插入靜音模式，插入靜音不佔用 SPI-Flash 記憶體的容量，一個位址位元可插入 10ms～25min 的靜音。

MP3 控制模式下，完全迎合市場上 MP3 的播放/暫停、停止、上一曲、下一曲、音量+、音量-等功能；

按鍵控制模式下觸發方式靈活，可隨意設置任意按鍵為脈衝可重複觸發、脈衝不可重複觸發、無效按鍵、電平保持不可迴圈、電平保持可迴圈、電平非保持可迴圈、上一曲不迴圈、下一曲不迴圈、上一曲可迴圈、下一曲可迴圈、音量+、音量-、播放/暫停、停止、播放/停止等 15 種觸發方式，最多可用 10 個按鍵觸發控制輸出。

3×8 按鍵組合控制模式下能以脈衝可重複觸發的方式觸發 24 個位址位元語音，所觸發位址位元語音可在 0～219 之間設置，本模組無法使用此模式。

並口控制模式可用 8 個控制埠進行控制，僅限於 WT588D-32L、WTW-28P，本模組無法使用此模式。

一線串口控制模式可通過發碼端控制語音播放、停止、迴圈播放和音量大小，或者直接觸發 0～219 位址位元的任意語音，發碼速度 600us～2000us。

三線串口控制模式和三線串口控制控制埠擴展輸出模式之間可通過發碼切換，三線串口控制模式下，能控制語音播放、停止、迴圈播放和音量大小，或者直接觸發 0～219 位址位元的任意語音，三線串口控制控制埠擴展輸出可以擴展輸出 8 位元，在兩種模式下切換，能讓上一個模式的最後一種狀態保持著進入下一個模式。

聲音輸出方式為 PWM 和 DAC 輸出方式，PWM 輸出可直接推動 0.5W/8Ω 的揚聲器，DAC 輸出外接功率放大器，音質非常好。

應用範圍廣，幾乎可以涉及到所有的語音場所，如報站器、報警器、提醒器、鬧鐘、學習機、智慧家電、治療儀、電子玩具、電訊、倒車雷達以及各種自動控制裝置等場所，工藝上達到工業應用的要求。

3

2.應用方塊圖

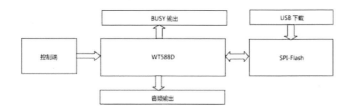

3.模組接腳圖

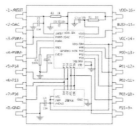

封裝引腳	引腳標號	簡述	功能描述
1	/RESET	/RESET	復位腳，低電平保持≥5ms 有效。
2	DAC	PWM+/DAC	PWM+/DAC 音訊輸出腳，視功能設置而定
3	PWM+	PWM+/DAC	PWM+/DAC 音訊輸出腳，視功能設置而定
4	PWM-	PWM-	PWM-音訊輸出腳
5	P14	DI	SPI-FLASH 通訊資料登錄腳
6	P13	DO	SPI-FLASH 通訊資料輸出腳
7	P16	CLK	SPI-FLASH 通訊時鐘腳
8	GND	GND	地線腳
9	P15	CS	SPI-FLASH 通訊片選腳
10	P03	K4/CLK/DATA	按鍵4/三線時鐘/一線資料登錄腳
11	P02	K3/CS	按鍵3/三線片選輸入腳
12	P01	K2/DATA	按鍵2/三線資料登錄腳
13	P00	K1	按鍵1
14	VCC	VDD-SIM	串口電源管理腳
15	P17	BUSY	語音播放忙信號輸出腳
16	VDD	VDD	電源輸入腳

4

● **MP3 模式**

MP3 模式下，WT588D 語音單片機自動預設 6 個控制埠的功能，對應列表如下：

控制埠	P00	P01	P02	P03
功能	停止	播放/暫停	下一曲	上一曲

● **按鍵控制模式**

所定義的腳位可以直接觸發晶片的一個功能，使晶片動作，每個腳位的觸發方式可單獨設置。按鍵控制模式的按鍵防抖時間為 10ms。

按鍵觸發模式下包括脈衝可重複觸發、脈衝不可重複觸發、電平保持可迴圈、電平保持不可迴圈、電平非保持迴圈、上一曲不迴圈、下一曲不迴圈、上一曲可迴圈、下一曲可迴圈、無效按鍵、播放/暫停、停止、音量+、音量-以及播放/停止等 15 種觸發方式。詳細控制方法見如下觸發時序圖。

1. 脈衝可重複觸發

　　負脈衝觸發。當控制埠檢測到有下降沿時（如該控制埠對地短路 25ms 以上），觸發播放語音。在語音播放期間，再檢測到下降沿，晶片會打斷正在播放的語音，重新播放。只要有下降沿信號，就重新播放。

2. 脈衝不可重複觸發

　　負脈衝觸發。當控制埠檢測到有下降沿時（如該控制埠對地短路 25ms 以上），觸發播放語音。在語音播放期間，再檢測到下降沿時，晶片不動作。直到語音結束後，檢測到的下降沿才有效。

3. 電平保持可迴圈

低電平觸發。當控制埠為低電平時，保持播放，高電平則停止。當第一遍播放結束後，還保持低電平，也不會繼續播放，觸發後只播放一次就結束。如果需要重新播放，則需要讓控制埠處於高電平，再拉為低電平，而後保持低電平即可。

4. 電平非保持迴圈

負脈衝/低電平觸發。當控制埠檢測到下降沿時（如該控制埠對地短路 25ms 以上），播放一遍語音然後停止。當控制埠為低電平時，保持播放，播放過程中，就算是給高電平也不停止，直到語音播放結束。當第一遍結束後，如果還保持低電平，則會繼續重複播放，只要不保持低電平且播放完當前語音後才停止。

5. 播放/停止

負脈衝觸發。控制埠檢測到下降沿時（如該控制埠對地短路 25ms 以上）開始播放，下一個下降沿結 音。不管聲音是束放處於播放還是停止狀態，都遵照這個規則。

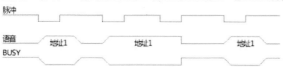

6. 下一曲不迴圈

負脈衝觸發。控制埠檢測到下降沿時（如該控制埠對地短路 25ms 以上）觸發播放下一段語音，下一個下降沿繼續觸發播放下一段，觸發播放完最後一段，則不會再有聲音。

6

7. 上一曲不迴圈

負脈衝觸發。控制埠檢測到下降沿時（如該控制埠對地短路 25ms 以上）觸發播放上一段語音，下一個下降沿繼續觸發播放上一段語音，播放完最前一段，則不再向前觸發播放語音。

8. 下一曲可迴圈

負脈衝觸發。控制埠檢測到下降沿時（如該控制埠對地短路 25ms 以上）觸發播放下一段語音，下一個下降沿繼續觸發播放下一段語音，重複操作，播放完最後一段語音，則會點播到第一段語音，如此迴圈觸發播放語音。

9. 上一曲可迴圈

負脈衝觸發。控制埠檢測到下降沿時（如該控制埠對地短路 25ms 以上）觸發播放上一段語音，下一個下降沿繼續觸發播放上一段語音，重複操作，播放完最前一段語音，則會點播到最後一段語音，如此迴圈觸發播放語音。

10. 暫停

負脈衝觸發。控制埠檢測到下降沿時（如該控制埠對地短路 25ms 以上）令正在播放的語音處於暫停狀態，下一個下降沿觸發暫停的語音從暫停點繼續播放。BUSY 在暫停狀態一直保持。

7

11. 停止

負脈衝觸發。控制埠檢測到下降沿時（如該控制埠對地短路 25ms 以上）令正在播放的語音停止。語音停止後再次觸發無效。

● **一線串口控制模式**

通過一根資料線發送串口資料。一線串口可以實現控制語音播放、停止、音量調節和直接觸發語音等功能。P00～P10 的按鍵可以選擇遮罩或者任意觸發方式。一線串口控制模式下，晶片無休眠狀態，語音停止後電流大約有 5mA，電池供電時請慎用。

1. 埠分配表

封裝形式	模組控制埠			
	P00	P01	P02	P03
WT588D-20SS	按鍵 K1	按鍵 K2	按鍵 K3	DATA

2. 命令及語音碼

命令碼	功能	描　　　述
E0H～E7H	音量調節	在語音播放或者待機狀態發此命令可以調節 8 級音量，E0H 最小，E7H 音量最大。
F2H	迴圈播放	在語音播放過程中發送此命令可迴圈播放當前位址語音。
FEH	停止語音播放	停止播放語音命令。

3. 語音位址對應關係

數據（十六進位）	功能
00H	播放第 0 段语音
01H	播放第 1 段语音
02H	播放第 2 段语音
……	……
D9H	播放第 217 段语音
DAH	播放第 218 段语音
DBH	播放第 219 段语音

4. 控制時序圖

一線串口只通過一條資料通信線控制時序，依照電平占空比不同來代表不同的資料位元。先發拉低 RESET 重定信號 5ms，然後置於高電平等待大於 17ms 的時間，再將資料信號拉低 5ms，最後發送資料。高電平與低電平資料占空比 1：3 即代表資料位元 0，高電平於低電平資料位元占空比為 3：1 代表數據位元 1。高電平在前，低電平在後。資料信號先發低位元再發高位。在發送資料時，無需先發送命令碼再發送指令。D0～D7 表示一個位址或者命令資料，資料中的 00H～DBH 為位址指令，E0H～E7H 為音量調節命令，F2H 為迴圈播放命令，FEH 為停止播放命令。詳細時序請見下圖：

說明：/RESET 為重定信號，發資料前對晶片進行重定，如不是在特殊的工業場合，可以不使用此重定信號。在每次發送資料前，不需要都發送重定信號，直接發送命令碼或者位址資料即可。DATA 為一線串口資料通信線，重定晶片穩定後先拉低 5ms 以喚醒晶片，低位元在前，BUSY 為語音晶片忙信號輸出，資料成功發送後等待 20ms，BUSY 輸出將作出回應。資料位元占空比對應詳見下圖。

| 200us | 600us | 高電平：低電平≈1：3，表示0 |
| 600us | 200us | 高電平：低電平≈3：1，表示1 |

5. 一線串口控制時序範例

例如，在一線串口控制模式下，發送資料 9CH 的時序參見下圖：

6. 範例程式

單晶片：PIC16F54，時脈：4MHz

```
Send oneline(unsigned char addr)
{
    rst=0;                      /*  對晶片進行重置    */
    delay1ms(5);                /*  重定信號保持低電平 5ms */
    rst=1;
    delay1ms(17);               /*  重定信號保持高電平 17ms */
    sda=0;
    delay1ms(5);                /*  資料信號置於低電平 5ms */
    for(i=0;i<8;i++)
    {
        sda=1;
        if(addr & 1)
        {
            delay1us(600);  /* 高電平比低電平為 600us：200us，表示發送資料 1 */
            sda=0;
            delay1us(200);
        }
        Else
        {
            delay1us(600);  /* 高電平比低電平為 200us：600us，表示發送資料 0 */
            sda=0;
            delay1us(200);
        }
        addr>>=1;
    }
    sda=1;
}
```

10

- **三線串口控制模式**

  三線串口控制模式由三條通信線組成，分別是片選 CS，資料 DATA，時鐘 CLK，時序根據標準 SPI 通信方式。通過三線串口可以實現語音晶片命令控制、語音播放。三線串口模式下，所有按鍵均無效。

  1. 埠分配表

封裝形式	模組控制埠			
	P00	P01	P02	P03
WT588D-20SS	———	DATA	CS	CLK

  2. 命令及語音碼

命令碼	功能	描　　　述
E0H～E7H	音量調節	在語音播放或者待機狀態發送此命令可以調節 8 級音量，E0H 最小，E7H 音量最大。
F2H	週圈播放	在語音播放過程中發送此命令可週圈播放當前位址語音。
FEH	停止語音播放	停止播放語音命令。
F5H	進入控制埠擴展輸出	在常規三線串口模式下，發送此命令可進入控制埠擴展輸出狀態。
F6H	退出控制埠擴展輸出	在控制埠擴展輸出狀態下，發送此命令可進入常規三線串口控制模式

  3. 語音位址對應關係

數據（十六進位）	功能
00H	播放第 0 段語音
01H	播放第 1 段語音
02H	播放第 2 段語音
……	……
D9H	播放第 217 段語音
DAH	播放第 218 段語音
DBH	播放第 219 段語音

  4. 三線串口控制時序

  三線串口控制模式由片選 CS、時鐘 CLK 和資料 DATA 腳組成，時序仿照標準 SPI 通信方式，重定信號在發碼前先拉低 5ms，然後拉高等待 17ms。工作時 RESET 需要一直保持高電平。片選信號 CS 拉低 5ms～10ms 以喚醒 WT588D 語音晶片，推薦使用 5ms。接收資料低位元在先，在時鐘的上升沿接收資料。時鐘週期介於 300us ～1ms 之間，推薦使用 300us。資料成功接收後，語音播放忙信號 BUSY 輸出在 20ms 之後做出回應。發資料時先發低位元，再發高位。發碼完成後 DATA、CLK、CS 均要保持高電平。在發送資料時，無需先發送命令碼再發送指令。D0～D7 表示一個位址或者命令資料，資料中的 00H～DBH 為位址指令，E0H～E7H 為音量調節命令，F2H 為迴圈播放命令，FEH 為停止播放命令，F5H 為進入三線串口控制控制埠擴展輸出命令，F6H 為退出三線串口控制控制埠擴
  展輸出命令。詳細時序圖如下：

11

說明：重定信號僅是在外因干擾比較強烈的環境中使用，如不是特殊的工業場合，不需要發送此重定信號，直接發送片選、時鐘和資料信號即可。

5. 命令碼發送時間

迴圈播放命令 F2H：迴圈播放命令需要在發送語音位址信號或者其他命令 30ms 之後、語音停止播放前發送，否則語音晶片不能有效接收。

停止播放命令 FEH：在語音播放的過程中發送此命令可以停止播放語音，在發送 DATA 信號 1ms 之後、語音停止前發送此命令則有效。

音量調節命令 E0H~E7H：在語音晶片工作狀態中發送此命令可以調節音量大小，不管語音晶片是否處於語音播放還是語音停止狀態。如果是先觸發位址語音或者其他的命令，則需要等待 90ms 才能發送音量調節命令，否則無效。

6. 範例程式

（主控單片機 PIC16F54，系統頻率 4MHz）

```
Send threelines(unsigned char addr)
{
    rst=0;                /*  對晶片進行重定  */
    delay1ms(5);          /*  重定信號保持低電平 5ms */
    rst=1;
    delay1ms(20);         /*  重定信號保持高電平 20ms */
    cs=0;
    delay1ms(5);          /*  片選信號保持低電平 5ms */
    for(i=0;i<8;i++)
    {
        scl=0;
        if(addr & 1)sda=1;
        else sda=0;
        addr>>=1;
        delay1us(150);    /*  時鐘週期 300us */
        scl=1;
        delay1us(150);
    }
        cs=1;
}
```

12

5.應用電路

按鍵控制電路

A. 按鍵控制 PWM 輸出應用電路

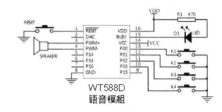

軟體設置：按鍵控制模式。

控制埠定義：選取控制埠 P00、P01、P02、P03 作為觸發埠，在燒錄軟體編輯 WT588D 語音專案時，把觸發埠的按鍵定義為可觸發播放的觸發方式，即可進行工作。

BUSY 輸出：工作信號輸出端，可從燒錄軟體端設置為播放狀態輸出為高電位或低電位。高電位時電壓接近 VDD 供電電壓，用于接 LED 做放音狀態指示或工作信號判斷。

供電電壓：VDD=DC2.8～5.5V，VCC=DC2.8～3.6V。採用 DC3.3V 供電時，可以直接短接 VDD 跟 VCC。採用 DC5V 供電時，VDD 端接 5V，VCC 端需要從 VDD 端串接兩個二極體以提供工作電壓。

音訊輸出：PWM 輸出方式，直接接喇叭。此種輸出方式下，PWM+、PWM-均不能短接到地或者接電阻電容到地。

B. 按鍵控制 DAC 輸出（外接電晶體）應用電路

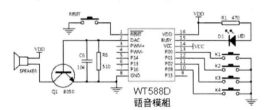

軟體設置、控制埠定義、BUSY 輸出與供電電壓設置皆與 PWM 輸出應用電路一樣。

音訊輸出：DAC 輸出方式，利用 NPN 電晶體將音訊信號放大再輸出給喇叭。R6 為分流電阻，設定值為 270Ω～1.2KΩ，阻值越大則輸出聲音越大。

13

C.按鍵控制 DAC 輸出（接功放）應用電路

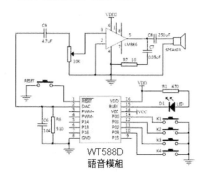

WT588D
語音模組

軟體設置、控制埠定義與 BUSY 輸出設置皆與 PWM 輸出應用電路一樣。

供電電壓：VDD=DC2.8～5.5V，VCC=DC2.8～3.6V。採用 DC3.3V 供電時，可以直接短接 VDD 跟 VCC，採用 DC5V 供電時，VDD 端接 5V，VCC 端需要從 VDD 端串接兩個二極體以提供工作電壓。VDD2 為功放工作電壓。

音訊輸出：DAC 輸出方式，音訊信號從 PWM+/DAC 端輸出，經過 R6、C6、C9 後輸出到功放。R6 為分流電阻，取值 270Ω～1.2KΩ，阻值越大則輸出聲音越大。

14

MCU 應用電路

一線串口控制 PWM 輸出應用電路

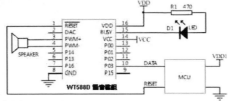

範例程式可參考 P17。

軟體設置：一線串口控制模式。

控制埠定義：P03 為 DATA 輸入腳，由 MCU 發送資料對 WT588D 語音模組進行控制。P00 ～P02 可以當作按鍵使用。

BUSY 輸出：工作信號輸出端，可從燒錄軟體端設置為播放狀態輸出為高電位或低電位。高 電位時電壓接近 VDD 供電電壓，用于接 LED 做放音狀態指示或工作信號判斷。

供電電壓：VDD=DC2.8～5.5V，VCC=DC2.8～3.6V。採用 DC3.3V 供電時，可以直接短接 VDD 跟 VCC；採用 DC5V 供電時，VDD 端接 5V，VCC 端需要從 VDD 端串接 兩個二極體以提供工作電壓。VDD1 為 MCU 工作電壓。如果 VDD1 跟 VDD 存在 壓差，則需要在 MCU 跟 WT588D 語音模組的通信線 DATA 上串接電阻。

音訊輸出：PWM 輸出方式，直接接揚聲器。此種輸出方式下，PWM+、PWM-均不能短接到 地或者接電阻電容到地。

三線串口控制 PWM 輸出應用電路

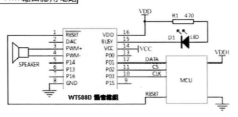

範例程式可參考 P18。

軟體設置：三線串口控制模式。

控制埠定義：P01 為 DATA，P02 為 CS，P03 為 CLK。由 MCU 發送資訊對模組進行控制。

供電電壓：VDD=DC2.8～5.5V，VCC=DC2.8～3.6V。採用 DC3.3V 供電時，可以直接短接 VDD 跟 VCC，採用 DC5V 供電時，VDD 端接 5V，VCC 端需要從 VDD 端串接 兩個二極體以提供工作電壓。VDD1 為 MCU 工作電壓。如果 VDD1 跟 VDD 存在 壓差，需要在 MCU 跟 WT588D 語音模組的通信線 DATA、CS、CLK 上串接電阻。

15

6.8051 範例程式

程式流程圖

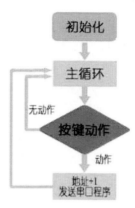

一線串口控制 C 語言程式

說明：此程式與 P15 的一線串口控制模式應用電路相互對應。
測試晶片：AT89S51。

```c
#include <AT89X51.H>
sbit KEY=P1^1;        /*P1_1 為 P1 的第 2 腳*/
sbit RST=P1^4;        /*P1_4 為 P3 的第 3 腳*/
sbit SDA=P3^0;        /*P3_0 為 P3 的第 4 腳*/
void delay1ms(unsigned char count)            //1MS 延時副程式
{
        unsigned char i,j,k;
        for(k=count;k>0;k--)
        for(i=2;i>0;i--)
        for(j=248;j>0;j--);
}
void delay100us(unsigned char count)          //100US 延時副程式
{
        unsigned char i;
        unsigned char j;
        for(i=count;i>0;i--)
        for(j=50;j>0;j--);
}
Send_oneline(unsigned char addr)
{
        unsigned char i;
        RST=0;
        delay1ms(5);            /*reset 延時 5MS*/
        RST=1;
        delay1ms(20);            /* delay 20ms */
        SDA=0;
        delay1ms(5);            /* delay 5ms */
        for(i=0;i<8;i++)
        {
                SDA=1;
                if(addr & 1)
                {
                        delay100us(6);      /* 600us */
                        SDA=0;
                        delay100us(2);      /* 200us */
                }
                else
                {
                        delay100us(2);      /* 200us */
                        SDA=0;
                        delay100us(6);      /* 600us */
                }
                addr>>=1; }
                SDA=1;
}
main()
{
        unsigned char FD=0;
        P3=0XFF;
        while(1)
        {
                if(KEY==0)
                {
                        delay1ms(10);
                        if(KEY==0)                     //通過按鍵 P1.1 來進行發碼值的遞增
                        {
                                Send_oneline(FD);
                                FD++;
                                if(FD==220) //一線串口時,總發碼段暫時數多為 220 段
                                {
                                        FD=0;
                                }
                                while(KEY==0);   //等待按鍵釋放,以免一次按鍵誤判成幾次
                        }
                }
        }
}
```

17

三線串口控制 C 語言程式

說明：此程式與 P15 的一線串口控制模式應用電路相互對應。

測試晶片：AT89S51 。

```c
#include <AT89X51.H>
sbit KEY=P1^1;      /*P1_1 為 P1 的第 2 腳*/
sbit RST=P1^4;      /*P1_4 為 P3 的第 3 腳*/
sbit CS=P3^1;       /*P3_1 為 P3 的第 4 腳*/
sbit SCL=P3^2;      /*P3_2 為 P3 的第 5 腳*/
sbit SDA=P3^0;      /*P3_0 為 P3 的第 6 腳*/
//sbit DENG=P3^7;   /*P3_5 為 P3 的第 6 腳*/
void delay1ms(unsigned char count) //1MS 延時副程式
{
    unsigned char i,j,k;
    for(k=count;k>0;k--)
    for(i=2;i>0;i--)
    for(j=248;j>0;j--);
}
void delay150us(void)      //150US 延時副程式
{
    unsigned char j;
    for(j=75;j>0;j--);
}
Send_threelines(unsigned char addr) //二線發傳副程式
{
    unsigned char i;
    RST=0;
    delay1ms(5);
    RST=1;
    delay1ms(20);          /*  reset 拉高 20ms*/
    CS=0;
    delay1ms(5);           /*  片選拉低 5ms */
    for(i=0;i<8;i++)
    {
        SCL=0;
        if(addr & 1)SDA=1;
        else SDA=0;
        addr>>=1;
        delay150us();      /* 150us */
        SCL=1;
        delay150us();
    }
    CS=1;
}
main()
{
    unsigned char FD=0;
    P3=0XFF;
    while(1)
    {
        if(KEY==0)
        {
            delay1ms(20);
            if(KEY==0)    //防篤過按鍵 P1.1 來進行發送值的調增
            {
                Send_threelines(FD);
                FD++;
                if(FD==220) //二線串口時,隨帶段點時最多為 220 段
                {
                    FD=0;
                }
                while(KEY==0);   //等待按鍵釋放,以免一次按鍵誤判成兩次
            }
        }
    }
}
```

18

WT588D 語音燒錄器操作手冊

WT588D 語音燒錄器

操作說明書

目　　錄

2

A. 軟體安裝

1. 點選進入 A22-0031 WT588D 語音燒錄器→燒錄軟體資料夾裡。

2. 點選 WT588D VoiceChip.exe

WT588D VoiceChip.exe

3. 使用 WIN7 的用戶在點選完會出現「使用者帳戶控制」的確認視窗，按下「是(Y)」確認。

4. 等待準備安裝的進度視窗。

5. 點選「 Next >」。

6. 進入使用協定說明，先點選「I accept the terms in the License agreement」，再點選「Next

3

＞」。

7. 先輸入使用著名稱(User Name)與企業名稱(Organization)，再選擇安裝使用對像為任何人(Anyone who uses this computer(all users))或只有自己使用(Only for me)，最後再點選「Next＞」。

8. 選擇安裝軟體的資料夾路徑，可點選「Change...」改變安裝路徑，選完後點選「Next ＞」。

9. 點選標準安裝(Typical)選項，然後再點選「Next >」。

10. 前面的選項都確認無誤後，點選「Install」開始安裝。

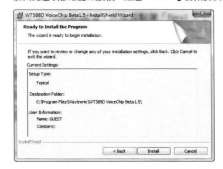

11. 開始安裝，稍待一會兒就會看到進度 bar 在移動，表示正在安裝。

5

12.安裝完成，點選「Finish」離開安裝程式。

B. 硬體介紹

以下為 WT588D 語音燒錄器各部圖解。

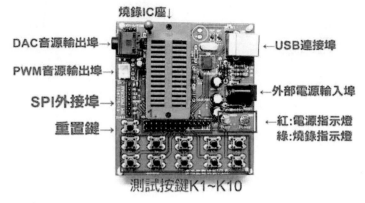

圖片說明：

- 紅框 1：為控制埠引出排針，上排為控制埠排針，從左到右依次為 P00、P01、P02、P03、P04、P05、P06、P07、P10、P11、P12（WT588D 語音模組只有用到 P00~P03）。下排全為 GND。

- DAC 音源輸出埠：為 3.5mm 圓孔座，在圓孔座上方位置有排針，使用 DAC 輸出時需將短路夾(jumper)接上，PWM 輸出時務必斷開。

- PWM 音源輸出埠：可外接喇叭。

- SPI 外接埠：提供線上下載輸出埠，把相關引線接到應用 WT588D 外掛 SPI-Flash 接腳上，再從電腦軟體上操作軟體進行下載即可。記憶體的接線部分如下圖。

Flash rom

SPI 輸出埠

- 測試按鍵 K1~K10：WT588D 語音模組只有用到 K1~K3。

- 外部電源輸入埠：可外接 DC5V~9V 的電源。

- 指示燈：電源輸入時紅色 LED 燈會亮；在燒錄時綠色指示燈會亮，平常不會亮。

- 燒錄 IC 座：燒錄時請注模組的擺放方向，模組的中央缺口方向一律朝上，底部底部對齊燒錄 IC 座最下方。如放反的話有可能會造成模組燒毀，請特別注意。

C. 操作步驟

1. 依照下圖將WT588D語音模組放到WT588D語音燒錄器的活動IC座上,注意! 將 WT588D語音模組的中央缺口方向朝上,底部對齊燒錄IC座最下方。確定後再將USB 線接上。

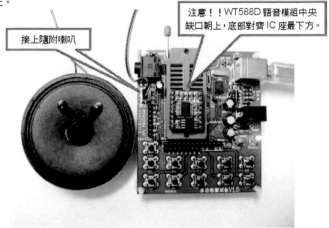

接上隨附喇叭

注意!!WT588D語音模組中央 缺口朝上,底部對齊IC座最下方。

2. 接上USB線後,電腦會偵測到新硬體,稍等一會兒驅動程式就會自動安裝完成。再到控 制台→系統→裝置管理員裡面,點開通用序列匯流控制器是否有USB Composite Device 裝置出,如果有,表示安裝完成。反之,請檢查USB是否有接好?或重新插入再試一試。

3. 開啟燒錄軟體，在開始功能表→所有程式→waytronic→WT588D VoiceChip Beta1.5。如下圖，一開始會介面上的文字有些是亂碼，選擇上方工具列的語言→ English，介面就會是變成英文畫面。

簡體中文介面(亂碼)　　　　　　　　　　英文介面(正常)

4. 選擇工具列的 File→NewProject，然後在另存新檔視窗裡選擇要儲存的資料夾位置，並輸入檔案名稱，然後按再按確定。

5. 在聲音檔載入區點選滑鼠右鍵→Load，選擇您已經錄好的 wav 檔或已轉好的 wav 檔案。使用者可自行上網下載免費錄音軟體或轉檔軟體，轉檔軟體建議可用 Freemake Audio Converter 進行 mp3 轉 wav 檔，裡面也可自行設定取樣率，在此就不多加敘述。

> 注意!如果要將一般 mp3 檔案轉 wav 檔時，記得先將檔案設為單音(MONO)、取樣率設定為6KHz、8KHz、12KHz、16KHz、22KHz，取樣大小設定為 16 bit。以上取樣率請皆設為整整，比如取樣率為 22.001KHz 時，則軟體將會警告無法正常讀入，請使用者多加注意。

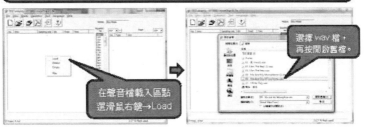

9

6. 重覆第 5 個步驟載入第二個 wav 檔，先選擇 Flash→8M，再選擇所要分段的位置，如下圖，放入第 1 段時則選擇 00H 的位置；放入第 2 段時則選擇 01H 的位置…以此類推。

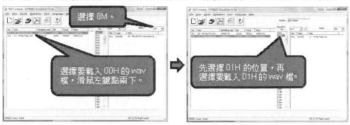

7. 如下圖點選 Compile 圖示，Compile 完成後會出現下圖右的視窗，表示成功了。如果出現 Fail 的視窗，表示可能聲音檔格式不對，請重新檢查。

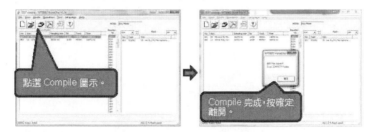

8. 如下圖選擇 Download 圖示後會出現下載視窗(如下圖右)，點選 Capacity→8M，當 Status 由紅色變為綠色，代表與語音模組連線成功。

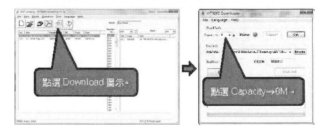

10

9. 點選 Erase 清除語音模組內部資料，清除完畢後再點 Download 的按鍵，下載完畢後 Status
會變回紅色圖示，此時語音就已經在語音模組內部了。

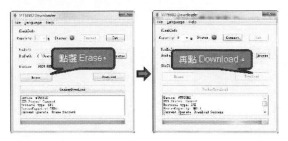

10. 回到語音燒錄的硬體進行測試，點下 K1 按鍵即可聽到第 1 段語音；點下 K2 按鍵即可聽
到第 2 段語音…以此類推。在這裡要注意一件事，因語音模組只支援 4 段語音分段，如有
4 段語音分段以上的需求，必須外接 MCU 控制分段位置，詳細情形請參考 WT588D 語音
模組的說明書。

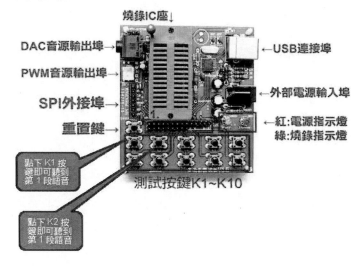

資料來源：燒錄操作說明書
http://img.web66.com.tw/_file/2307/upload/download/diskdata/A22-0031WT588D%E8%AA
%9E%E9%9F%B3%E7%87%92/WT588D%E8%AA%9E%E9%9F%B3%E7%87%92%E9
%8C%84%E5%99%A8%E8%AA%AA%E6%98%8E%E6%9B%B8.pdf

WT588D 語音燒錄器操作手冊(英文版)

WT588D SOFTWARE USER MANUAL

Catalog

1. SOFTWARE INSTALL

Double click the "setup_E.exe" start to setup the software to PC.

Setup.WT_App.msi   Setup.WT...   setup_C.exe   setup_E.exe

2. SOFTWARE OPERATIONS

2.1. SOFTWARE INTERFACE

Voice loading area: Load voice files here ,Voice editing area: Insert voice files form loading area, and edit. Voice files in editing area will not take up any memory space.

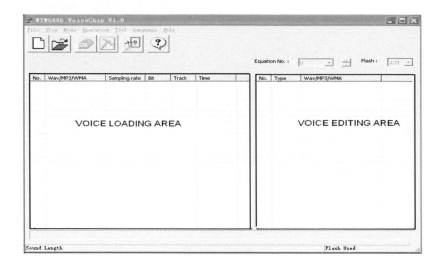

2.1.1. KEY INFORMATIONS

They are "New project", "Open project", "Compile (save BIN)", "Options", "Download data".

2.2. NEW PROJECT

Click "File" →"New project" to create a new project.

Name the project in the pop-up dialog box, and choose a folder to save it.

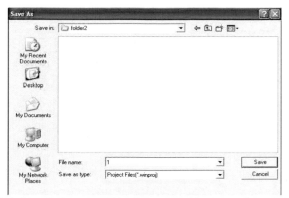

Press "save" the project will save in a new project folder, this system new folder include all the project information.

2.3. MEMORY SIZE SETTINGS

In the top right corner of software interface, SPI-Flash size can be set. There are 2M, 4M, 8M, 16M, 32M, 64M options. Choose the size according to your need. Please refer to "**WT588D SUPPLY INFORMATION**".

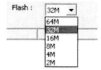

2.4. LOADING VOICE

Voice sampling rate must be 6000Hz、8000Hz、10000Hz、12000Hz、14000Hz、16000Hz、18000Hz or 20000Hz, right click the mouse button and load the voice files.

3

Choose the voice files, and open.

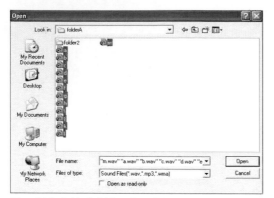

After voice loaded in, their information will show up, such as file name, sampling rate, bit, track, duration.

No.	Wav/MP3/WMA	Sampling rate	Bit	Track	Time
001	a.wav	12000 Hz	16 Bit	MONO	8802 ms
002	b.wav	12000 Hz	16 Bit	MONO	8840 ms
003	c.wav	12000 Hz	16 Bit	MONO	8411 ms
004	d.wav	12000 Hz	16 Bit	MONO	4060 ms
005	e.wav	12000 Hz	16 Bit	MONO	9606 ms
006	f.wav	12000 Hz	16 Bit	MONO	6198 ms
007	g.wav	10000 Hz	16 Bit	MONO	1262 ms
008	h.wav	10000 Hz	16 Bit	MONO	1029 ms
009	i.wav	10000 Hz	16 Bit	MONO	921 ms
010	j.wav	10000 Hz	16 Bit	MONO	910 ms
011	k.wav	10000 Hz	16 Bit	MONO	813 ms
012	l.wav	10000 Hz	16 Bit	MONO	1048 ms
013	m.wav	10000 Hz	16 Bit	MONO	895 ms

2.5. MODES OPTIONS.

Click "Operation" →" Options" enter into the interface.

4

There are 6 control modes, default is "Key mode".

2.5.1. ONE LINE SERIAL MODE

Click "Operation" →"Options", choose "one line mode" and press "OK".

Click "Operation" →"Key setup", you can see each I/O corresponding KEY, KEY 1 to KEY 10 default. KEY 4 (corresponding I/O P03) is locked up, P03 is DATA pin in the mode, can not be used as a key. Other I/Os can be used as keys. Trigger mode including "No function", "Edge retrigger", "Edge no retrigger", "Level hold loop", "Level loop" ,"On/off" ,"Next unloop", "Prev unloop" ,"Next loop" ," Prev loop" ,"Level unloop", "Pause" ,"Vol+", "Vol-","Stop".

After choose trigger mode, when the corresponding key(I/O) set to low level, voice can be trigger to play in the set trigger mode.

The available keys(I/Os) were defaulted as "Edge retrigger".

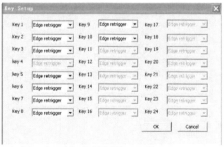

Click "Operation" →"Equation setup", to set each key's (I/O's) trigger address.

5

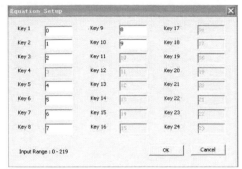

KEY 1 to KEY 10 default trigger address 0~9, the addresses can be changed input number between 0~219. The following trigger modes can direct trigger voice addresses . "Edge retrigger", " Edge no retrigger" ,"Level hold loop", "Level loop"," On/off" ,"Level unloop". Only the corresponding key(I/O) set to these trigger modes, addresses can be triggered directly.

Addresses information please refer to **2.8 VOICE ADDRESSES**

2.5.2. THREE LINE SERIAL MODE

Click "Operation" →"Options", choose "three line mode" and press "OK". In this mode, all the I/Os can not be used as keys. As you can see all the keys were locked up.

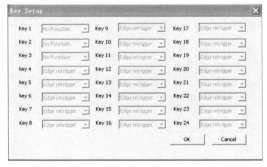

Open "Operation" →"Equation" ,Key 2(P01), Key 3(P02), Key4(P03) were locked up. Other keys also invalid here.

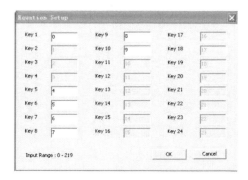

2.5.3. MP3 MODE

Click "Operation" →"Options", choose "MP3 mode" and press "OK".

In this mode, Key 1 to Key 6 default as "STOP", "ON/OFF","NEXT","PREV","VOL+","VOL-" individually. Other I/Os are invalid.

Open "Operation" →"Equation", Key 1 to key 6 addresses were locked up. Others invalid.

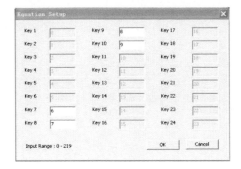

2.5.4. KEY MODE

Click "Operation" →"Options", choose "MP3 mode" and press "OK".

In this mode, Key 1(P00), Key2(P01), Key3(P02), Key4(P03), Key5(P04), Key6(P05), Key7(P06), Key8(P07), Key9(P10), Key 10(P11) all are valid. Open "Operation" →"Key set", Keys (I/Os) trigger mode can be changed by drop-down menu. The default trigger mode is Edge retrigger.

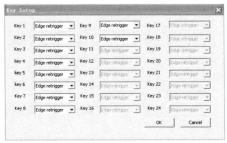

Trigger addresses can be changed by input addresses.

8

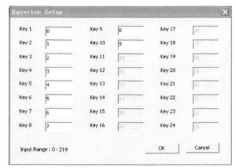

KEY 1 to KEY 10 default trigger address 0~9, the addresses can be changed input number between 0~219. The following trigger modes can direct trigger voice addresses . "Edge retrigger", " Edge no retrigger" ,"Level hold loop", "Level loop"," On/off" ,"Level unloop". Only the corresponding key(I/O) set to these trigger modes, addresses can be triggered directly.

Addresses information please refer to **2.8 VOICE ADDRESSES**

2.5.5. MATRIX 3x8 MODE

Click "Operation" →"Options", choose "Matrix 3X8 mode" and press "OK".

In this mode, voice address triggered by Matrix(consist of I/Os), all the keys were locked up, and Edge retrigger. Open "Operation" →"Key setup" , all keys locked.

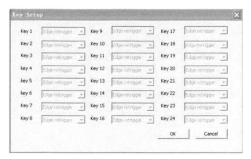

Open "Operation" →"Equation setup", 24 voice addresses can be set from 0~219

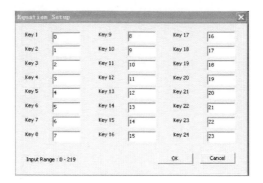

2.5.6. PARALLEL MODE

Click "Operation" →"Options", choose "Matrix 3X8 mode" and press "OK".

In this mode, Key1 (P00) defined as SBT pin, P01, P02, P03, P04, P05, P06, P07, P10 are addresses.

Open "Operation" →"Key setup", only key 1's trigger mode can be changed. We suggest set to "Edge retrigger" or " Edge no retrigger" or "Level hold loop" or "Level loop" or " On/off" or "Level unloop". Because these modes can direct trigger voice addresses.

Open "Operation" →"Key setup", key1 setup change is invalid.

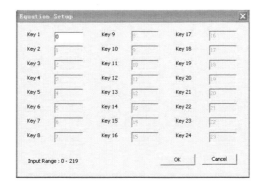

2.6. AUDIO OUTPUT

Open "Operation" →"Options" and choose audio output mode.
DAC output: external amplifier is needed
PWM output: direct drives speaker

2.7. BUSY SETTING

Open "Operation" →"Options", to set the BUSY port(I/O P17) high level or low level when playing voice.

2.8. VOICE ADDRESSES

Choose the Equation NO. by "+" or "-" ,and load (double click left button or click right button)the voice files from "VOICE LOADING AREA", voice files can be reused.

11

There are 0~219 options in " Equation NO.", up addresses by "+" (or "W") , down addresses by "-"(or "S") .

For example, set Key 1 trigger mode as "Edge retrigger", Equation No. as 0, Key 1(P00) can trigger voices to play in order in address 0.

There are 220 voice addresses total. 85 groups of voice can be load to each address max. Mute also can be inserted. The same voice use in other addresses, will not take additional memory space. (just take one voice space). At the same time, mute will not take space also. The total size depends on the size of voices in "VOICE LOADING AREA".

2.9. INSERT MUTE (SILENCE)

Right click mouse button in "VOICE EDITING AREA", choose "Insert" → " User defined silence ", also 10ms、20ms、50ms、100ms、200ms、300ms silence can be chosen directly.

For example insert 10ms mute between voice 1 and voice 2 , after voice 1 played, there are 10ms mute, and then play voice 2.

Also in "User defined silence" insert silence. Such as insert 100ms.

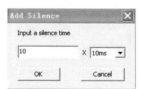

After input multiple, and "OK" 100ms silence already inserted.

No.	Type	Wav/MP3/WMA	
001	SOUND	a.wav	
002	SOUND	b.wav	
003	SILENCE	100 ms	
004	SOUND	c.wav	
005	SOUND	d.wav	

2.10. DELETE AND EMPTY VOICES

In the "VOICE LOADING AREA" we can choose the voice and delete one by one. Also choose "EMPTY", to delete all the voices by one time . also will empty all the voice in" VOICE EDITING AREA".

In the" VOICE EDITING AREA", we can choose the voice and delete one by one. Choose "EMPTY", to delete all the voices in current address by one time. Other addresses voices will not be deleted.

2.11. COMPILE (SAVE BIN)

After finish the project, we have to compile it as BIN file, and download to SPI-Flash on WT588D module.

Open "Operation" →"Compile (save bin)" or press"F4", to compile Bin file.

If the Bin file size big than the SPI-Flash memory, there will be a pop up dialog box to inform you change bigger memory. The percentage (the Bin file take the memory size) will show in the lower right corner .

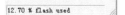

2.12. DOWNLOAD DATA

Put the module on the programmer properly and connect to the PC by USB ,Open "Tool" →"download data", or press F5 ,download data to Flash memory.

Select the Flash size you are going to program. And then click "Connect" .

Click "OK" to keep going.

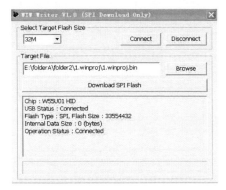

Click "Download SPI Flash", and will into erasing status.

After finish erasing, into Downloading status.

14

Download Ok, all the data already in SPI –Flash memory.

15

2.13. OPERATION STEPS

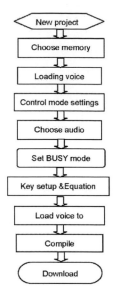

Note: If you already have Bin file, you can open it by PC software and download data directly, Unnecessary to editing the voice again.

Serial MP3 Player 參考手冊

1 Description

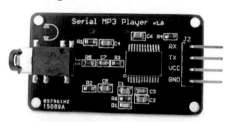

The module is a kind of simple MP3 player device which is based on a high-quality MP3 audio chip---YX5300. It can support 8k Hz ~ 48k Hz sampling frequency MP3 and WAV file formats. There is a TF card socket on board, so you can plug the micro SD card that stores audio files. MCU can control the MP3 playback state by sending commands to the module via UART port, such as switch songs, change the volume and play mode and so on. You can also debug the module via USB to UART module. It is compatible with Arduino / AVR / ARM / PIC.

Features:

1. Support sampling frequency (kHz): 8 / 11.025 / 12 / 16 / 22.05 / 24 / 32 / 44.1 / 48
2. High quality
3. Support file format: MP3 / WAV
4. Support Micro SD card, Micro SDHC Card
5. 30 class adjustable volume
6. UART TTL serial control playback mode, baud rate is 9600bps
7. Power supply can be 3.2 ~ 5.2VDC
8. Control logic interface can be 3.3V / 5V TTL
9. Compatible with Arduino UNO / Leonardo / Mega2560 / DUE

2 Specification

Item	Min	Typical	Max	Unit
Power Supply(VCC)	3.2	**5**	5.2	VDC
Current (@VCC=5V)	/	/	200	mA
Logic interface	3.3V / 5V TTL			/
Supported Card Type	Micro SD card(<=2G); Mirco SDHC card(<=32G)			/
File system format	Fat16 / Fat32			/
Uart baud rate	**9600**			bps
Dimensions	49X24X8.5			mm
Net Weight	5			g

3 Interface

4x M2 mounting holes   TF card socket

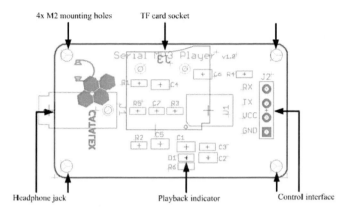

Headphone jack   Playback indicator   Control interface

Control interface: It is UART TTL interface. A total of four pins (GND, VCC, TX, RX), GND to ground, VCC is the power supply, TX is the TX pin of the MP3 chip, RX is the RX pin of the MP3 chip.

TF card socket: The micro sd card can be plugged in it.

Playbck indicator: Green light. If it is ready to play or it is paused, it keeps lighting. If playing, it blinks.

Headphone jack: It can be connected with the headphone or external amplifier.

Mounting holes: 4 screw mounting holes whose diameter is 2.2mm, so that the module is easy to install, easy to combine with other modules.

4 Usage

4.1 About the commands

4.1.1 Asynchronous serial port control play mode:

<table>
<tr><th colspan="3">Command bytes: SS VER Len CMD Feedback data SO</th></tr>
<tr><th>Mark</th><th>Byte</th><th>Byte description</th></tr>
<tr><td>SS</td><td>0x7E</td><td>Every command should start with S(0x7E)</td></tr>
<tr><td>VER</td><td>0xFF</td><td>Version information</td></tr>
<tr><td>Len</td><td>0xxx</td><td>The number of bytes of the command without starting byte and ending byte</td></tr>
<tr><td>CMD</td><td>0xxx</td><td>Such as PLAY and PAUSE and so on</td></tr>
<tr><td>Feedback</td><td>0xxx</td><td>0x00 = not feedback, 0x01 = feedback</td></tr>
<tr><td>data</td><td></td><td>The length of the data is not limit and usually it has two bytes</td></tr>
<tr><td>SO</td><td>0xEF</td><td>Ending byte of the command</td></tr>
</table>

4.1.2 Commonly Command bytes Descriptions:

Command	Command bytes without checksum(HEX)	Remark
[Next Song]	7E FF 06 **01** 00 00 00 EF	
[Previous Song]	7E FF 06 **02** 00 00 00 EF	
[Play with index]	7E FF 06 **03** 00 **00 01** EF	Play the first song
	7E FF 06 **03** 00 **00 02** EF	Play the second song
[Volume up]	7E FF 06 **04** 00 00 00 EF	Volume increased one
[Volume down]	7E FF 06 **05** 00 00 00 EF	Volume decrease one
[Set volume]	7E FF 06 **06** 00 00 **1E** EF	Set the volume to 30 (0x1E is 30)
[Single cycle play]	7E FF 06 **08** 00 00 **01** EF	Single cycle play the first song
[Select device]	7E FF 06 **09** 00 00 **02** EF	Select storage device to TF card
[Sleep mode]	7E FF 06 **0A** 00 00 00 EF	Chip enters sleep mode
[Wake up]	7E FF 06 **0B** 00 00 00 EF	Chip wakes up
[Reset]	7E FF 06 **0C** 00 00 00 EF	Chip reset
[Play]	7E FF 06 **0D** 00 00 00 EF	Resume playback
[Pause]	7E FF 06 **0E** 00 00 00 EF	Playback is paused
[Play with folder and file name]	7E FF 06 **0F** 00 **01 01** EF	Play the song with the directory: /01/001xxx.mp3
	7E FF 06 **0F** 00 **01 02** EF	Play the song with the directory: /01/002xxx.mp3
[Stop play]	7E FF 06 **16** 00 00 00 EF	
[Cycle play with folder name]	7E FF 06 **17** 00 **01 02** EF	01 folder cycle play
[Set single cycle play]	7E FF 06 **19** 00 00 **00** EF	Start up single cycle play
	7E FF 06 **19** 00 00 **01** EF	Close single cycle play

[Set DAC]	7E FF 06 1A 00 00 00 EF	Start up DAC output
	7E FF 06 1A 00 00 01 EF	DAC no output
[Play with volume]	7E FF 06 22 00 1E 01 EF	Set the volume to 30 (0x1E is 30) and play the first song
	7E FF 06 22 00 0F 02 EF	Set the volume to 15(0x0f is 15) and play the second song

4.2 Use USB to Uart TTL module

(1) You need a **USB to Uart TTL module** (such as USB/Serial Adapter) to connect **Serial MP3 Player** to PC. The hardware installation as show below:

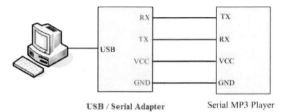

USB / Serial Adapter   Serial MP3 Player

(2) After the connection is completed, open the sscom32 serial tool that you can down load from catalex net disk to send commands. About the specific commands, please refer to 4.1.1 part.

(3) Click the EXT button and then you can manage the commands to be sent.

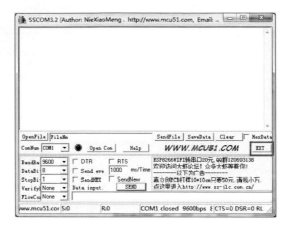

(3) Baud rate should be 9600. Tick HEX and HexData so that the command can be received by the Serial MP3 Player and you can see the feedback information (refer to the file **YX5300-24SS Datasheet V1.0.pdf**) in the blank of the window. Before sending commands, you should select the [ComNum] and click [Open Com].

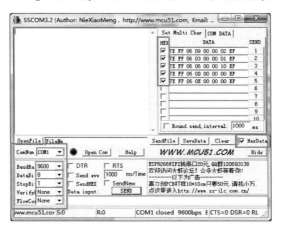

(4) Make sure your micro sd card is formatted as FAT16 or FAT32 and there is some songs in it. May be you should creat folder "01" and "02" , and put some songs with the name 001xxx.mp3 / 002xxx.mp3 / 003xxx.mp3 in the two folder. Some commands need them.

(5) After power up, you should send the command [Select device] first. Serial MP3 Player only supports micro sd card, so you should send " 7E FF 06 09 00 00 02 EF " .

Then you can send the command [**Play with index**] to play some song.

You can send the command [**Set volume**] to set the volume(0 ~ 30 class).

More operations ? Please refer to 4.1.1 part.

4.3 Use Arduino UNO R3

4.3.1   Project1: Simple test for the player.

Step1: Material preparation
   1x Arduino UNO R3
   1x USB Cable
   1x Serial MP3 Player
   1x Base Shield
   4x Female to Female Dupont cables

Step2: Hardware install
   (1)Plug the Base Shield which is just the I/O expansion board to Arduino UNO R3.
   (2)Connect the modules and Base Shield with the cables:

Serial MP3 Player	Wire	Base Shield
GND	<--->	GND
VCC	<--->	5V
TX	<--->	D5
RX	<--->	D6

   (3)Make sure your micro sd card is formatted as FAT16 or FAT32 and there is some songs in it. May be you should creat folder "01" and "02", and put some songs with the name 001xxx.mp3 / 002xxx.mp3 / 003xxx.mp3 in the two folder. Some commands need them.
Plug the micro sd card into the TF card socket on the Serial MP3 Player, and then plug the headphone.

Step3: Power on

Use the USB cable to connect the Arduino UNO R3 and PC.

Step4: Upload the demo code

Download the demo code (SerialMP3PlayerDemoforArduino-1.0.zip), and unzip it to your code project folder such as ../Arduino-1.0/MyProject. And then upload the code to your arduino UNO R3.

Step5: Enjoy yourself

Push the reset button on the Base Shield to play the first song in the micro sd card.

Step6: Power off

Unplug USB cable.

4.3.2 Project2: Use some modules to control the player. Enjoy!

Step1: Material preparation

1x Arduino UNO R3

1x USB Cable

1x Serial MP3 Player

1x Base Shield

1x Touch Sensor

1x Rotary Angle Sensor

10x Female to Female Dupont cables

Step2: Hardware install

(1)Plug the Base Shield which is just the I/O expansion board to Arduino UNO R3.

(2)Connect the modules and Base Shield with the cables:

	Wire	Base Shield
GND	<--->	GND
VCC	<--->	5V
TX	<--->	D5
RX	<--->	D6

Touch Sensor	Wire	Base Shield
GND	<--->	GND
VCC	<--->	5V
SIG	<--->	D2

Rotary Angle Sensor	Wire	Base Shield
GND	<--->	GND
VCC	<--->	5V
SIG	<--->	A0

(3)Make sure your micro sd card is formatted as FAT16 or FAT32 and there is some songs in it. May be you should creat folder "01" and "02", and put some songs with the name 001xxx mp3 / 002xxx.mp3 / 003xxx.mp3 in the two folder. Some commands need them.

Plug the micro sd card into the TF card socket on the Serial MP3 Player, and then plug the headphone.

Step3: Power on

Use the USB cable to connect the Arduino UNO R3 and PC.

Step4: Upload the demo code. If you have download in Project1, skip this step.

Download the demo code (SerialMP3PlayerDemoforArduino-1.0.zip), and unzip it to your code project folder such as ../Arduino-1.0/MyProject. And then upload the code to your arduino UNO R3.

Step5: Enjoy yourself

Push the reset button on the Base Shield. In the process that the Rotation Angle Sensor is rotated from the 'Min' side to the 'Max' side, the volume is gradually greater. If you touch the Touch Sensor,it will play or pause.

About more specific commands, please refer to **4.1.1** part.

Step6: Power off

Unplug USB cable.

DFPLayer Mini 参考手册

1. Summary

1.1 .Brief Instruction

DFPLayer Mini module is a serial MP3 module provides the perfect integrated MP3, WMV hardware decoding. While the software supports TF card driver, supports FAT16, FAT32 file system. Through simple serial commands to specify music playing, as well as how to play music and other functions, without the cumbersome underlying operating, easy to use, stable and reliable are the most important features of this module.

1.2 .Features

➢ Support Mp3 and WMV decoding
➢ Support sampling rate of
8KHz,11.025KHz,12KHz,16KHz,22.05KHz,24KHz,32KHz,44.1KHz,48KHz
➢ 24-bit DAC output, dynamic range support 90dB, SNR supports 85dB
➢ Supports FAT16, FAT32 file system, maximum support 32GB TF card
➢ A variety of control modes, serial mode, AD key control mode
➢ The broadcast language spots feature, you can pause the background music being played
➢ Built-in 3W amplifier
➢ The audio data is sorted by folder; supports up to 100 folders, each folder can be assigned to 1000 songs
➢ 30 levels volume adjustable, 10 levels EQ adjustable.

1.3 .Application

➢ Car navigation voice broadcast
➢ Road transport inspectors, toll stations voice prompts
➢ Railway station, bus safety inspection voice prompts
➢ Electricity, communications, financial business hall voice prompts
➢ Vehicle into and out of the channel verify that the voice prompts
➢ The public security border control channel voice prompts
➢ Multi-channel voice alarm or equipment operating guide voice
➢ The electric tourist car safe driving voice notices
➢ Electromechanical equipment failure alarm
➢ Fire alarm voice prompts

➢ The automatic broadcast equipment, regular broadcast.

2. Module Application Instruction

2.1. Specification Description

Item	Description
MP3Format	1、Support 11172-3 and ISO13813-3 layer3 audio decoding
	2、Support sampling rate (KHZ):8/11.025/12/16/22.05/24/32/44.1/48
	3、Support Normal、Jazz、Classic、Pop、Rock etc
UART Port	Standard Serial; TTL Level; Baud rate adjustable(default baud rate is 9600)
Working Voltage	DC3.2~5.0V; Type :DC4.2V
Standby Current	20mA
Operating Temperature	-40~+70
Humidity	5% ~95%

Table 2.1 Specification Description

2.2 .Pin Description

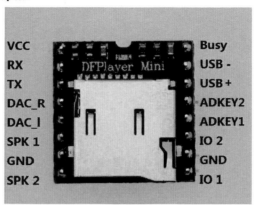

Figure 2.1

No	Pin	Description	Note
1	VCC	Input Voltage	DC3.2~5.0V;Type: DC4.2V
2	RX	UART serial input	
3	TX	UART serial output	
4	DAC_R	Audio output right channel	Drive earphone and amplifier
5	DAC_L	Audio output left channel	Drive earphone and amplifier.
6	SPK2	Speaker	Drive speaker less than 3W
7	GND	Ground	Power GND
8	SPK1	Speaker	Drive speaker less than 3W
9	IO1	Trigger port 1	Short press to play previous（long press to decrease volume）
10	GND	Ground	Power GND
11	IO2	Trigger port 2	Short press to play next（long press to increase volume）
12	ADKEY1	AD Port 1	Trigger play first segment
13	ADKEY2	AD Port 2	Trigger play fifth segment
14	USB+	USB+ DP	USB Port
15	USB-	USB- DM	USB Port
16	BUSY	Playing Status	Low means playing \High means no

Table 2.2 Pin Description

3. Serial Communication Protocol

Serial port as a common communication in the industrial control field, we conducted an industrial level of optimization, adding frame checksum, retransmission, error handling, and other measures to significantly strengthen the stability and reliability of communication, and can expansion more powerful RS485 for networking functions on this basis, serial communication baud rate can set as your own, the default baud rate is 9600

3.1. Serial Communication Format

Support for asynchronous serial communication mode via PC serial sending commands
Communication Standard:9600 bps
Data bits :1
Checkout :none
Flow Control :none

Format: SS VER Len CMD Feedback para1 para2 checksum SO		
$S	Start byte 0x7E	Each command feedback begin with $, that is 0x7E
VER	Version	Version Information
Len	the number of bytes after "Len"	Checksums are not counted
CMD	Commands	Indicate the specific operations, such as play / pause, etc.
Feedback	Command feedback	If need for feedback, 1: feedback, 0: no feedback
para1	Parameter 1	Query high data byte
para2	Parameter 2	Query low data byte
checksum	Checksum	Accumulation and verification [not include start bit $]
$O	End bit	End bit 0xEF

For example, if we specify play NORFLASH, you need to send: 7E FF 06 09 00 00 04 FF DD EF
Data length is 6, which are 6 bytes [FF 06 09 00 00 04]. Not counting the start, end, and verification.

3.2 .Serial Communication Commands

1).Directly send commands, no parameters returned

CMD	Function Description	Parameters(16 bit)
0x01	Next	
0x02	Previous	
0x03	Specify tracking(NUM)	0-2999
0x04	Increase volume	
0x05	Decrease volume	
0x06	Specify volume	0-30
0x07	Specify EQ(0/1/2/3/4/5)	Normal/Pop/Rock/Jazz/Classic/Base
0x08	Specify playback mode (0/1/2/3)	Repeat/folder repeat/single repeat/ random

0x09	Specify playback source(0/1/2/3/4)	U/TF/AUX/SLEEP/FLASH
0x0A	Enter into standby – low power loss	
0x0B	Normal working	
0x0C	Reset module	
0x0D	Playback	
0x0E	Pause	
0x0F	Specify folder to playback	1~10(need to set by user)
0x10	Volume adjust set	{DH = 1:Open volume adjust }{DL: set volume gain 0~31}
0x11	Repeat play	{1:start repeat play}{0:stop play}

2).Query the System Parameters

Commands	Function Description	Parameters(16 bit)
0x3C	STAY	
0x3D	STAY	
0x3E	STAY	
0x3F	Send initialization parameters	0 - 0x0F(each bit represent one device of the low-four bits)
0x40	Returns an error, request retransmission	
0x41	Reply	
0x42	Query the current status	
0x43	Query the current volume	
0x44	Query the current EQ	
0x45	Query the current playback mode	
0x46	Query the current software version	
0x47	Query the total number of TF card files	
0x48	Query the total number of U-disk files	
0x49	Query the total number of flash files	
0x4A	Keep on	
0x4B	Queries the current track of TF card	
0x4C	Queries the current track of U-Disk	
0x4D	Queries the current track of Flash	

3.3. Returned Data of Module

3.3.1. Returned Data of Module Power-on

1).The module power on, require a certain of the time initialization, this time is determined by U-disk, TF card, flash, etc. device 's file numbers, general situation in the 1.5 ~ 3Sec. If module initialization data has not been sent out within the time, indicating that the module initialization error, please reset the module's power supply, and detect hardware connecting;

2).The module initialization data including online devices, such as sending 7E FF 06 3F 00 00 01 xx xx EF, DL = 0x01 describe only the U-disk online during power-on, Other data are seen as the table below:

U-Disk on-line	7E FF 06 3F 00 00 01 xx xx EF	Each device are or relationship
TF Card on-line	7E FF 06 3F 00 00 02 xx xx EF	
PC on-line	7E FF 06 3F 00 00 04 xx xx EF	
FLASH on-line	7E FF 06 3F 00 00 08 xx xx EF	
U-disk & TF Card on-line	7E FF 06 3F 00 00 03 xx xx EF	

3).MCU will not send corresponding control commands until module initialization sending commands or the module will not process the commands sent by MCU, and will also affect the normal initialization of the module.

3.3.2 .Returned Data of Track Finished Playing

U-Disk finish playback 1st track	7E FF 06 3C 00 00 01 xx xx EF
U-Disk finish playback 2nd track	7E FF 06 3C 00 00 02 xx xx EF
TF card finish playback 1st track	7E FF 06 3D 00 00 01 xx xx EF
TF card finish playback 2nd track	7E FF 06 3D 00 00 02 xx xx EF
Flash finish playback 1st track	7E FF 06 3E 00 00 01 xx xx EF
Flash finish playback 2nd track	7E FF 06 3E 00 00 02 xx xx EF

1.The module will enter into pause status automatically after being specified playing, if customers need such application, they can specify track to play ,the module will enter into pause status after finishing playing ,and wait for the commands sent by MCU.

2 In addition, we opened a dedicated I/O as decoding and pausing status indication. See Pin 16, Busy
1).Output high level at playback status;
2).Output low level at pause status and module sleep;

3. For continuous playback applications, it can be achieved as below, if it finishes the first tracking of the TF card, it will return

7E FF 06 3D 00 00 01 xx xx EF
3D ---- U-disk command
00 01 ---- expressed finished playing tracks.

If the external MCU receives this command, please wait 100ms. And then sending the playback command [7E FF 06 0D 00 00 00 FF EE EF], because inside the module it will first initialize the next track information. In this case, the module can be played continuously.

4. If the currently finish playing the first song, the track pointer automatically point to second song, If you send a "play the next one" command, then the module will playback the third song. And, if the module finishes playing the last one, the player will automatically jump to the first pointer, and pause.

5. After specifying device, the module play pointer will point to device root directory of the first track, and enters the pause state, and wait MCU sending track playing command.

3.3.3 .Returned Data of Module Responds

FLASH finish play the 1st track	7E FF 06 3E 00 00 01 xx xx EF

1). in order to strengthen the stability of the data communication, we have increased response processing; ACKB byte is set whether need to reply to response. So that to ensure each communication get handshake signals, which will indicate the module has been successfully received data sent by the MCU and process immediately.

2).For general applications, customers can freely choose, without this response processing is also ok.

3.3.4 .Returned Data of Module Error

Module is busy	7E FF 06 40 00 00 00 xx xx EF
A frame data are not all received	7E FF 06 40 00 00 01 xx xx EF
Verification error	7E FF 06 40 00 00 02 xx xx EF

1). In order to strengthen the stability of the data communication, we added data error handling mechanism. Module will responds information after receiving error data format;

2). In the case of relatively harsh environment, it is strongly recommended that customers process this command. If the application environment in general, you no need handle it;

3).The module returns busy, basically when module power-on initialization will return, because the modules need to initialize the file system.

3.3.5. Push-in and Pull-out information of Device

Push in U-disk	7E FF 06 3A 00 00 01 xx xx EF
Push in TF card	7E FF 06 3A 00 00 02 xx xx EF
Pull out U-disk	7E FF 06 3B 00 00 01 xx xx EF
Pull out TF card	7E FF 06 3B 00 00 02 xx xx EF

1).For the flexibility of the module, we particularly add command feedback of push-in and pull-out device. Let user know the working status of the module.

2).When push-in device, we default playback the first track of device root directory as audition, if users do not need this feature, you can wait 100ms after receiving the message of push –in serial device ,and then send pause command.

3.4 Serial Commands

3.4.1. Commands of Specify Track Play

Our instructions are given in support of the specified track is playing, the song selection ranges from 0 to 2999. Actually can support more, because it involves the reasons to the file system, support for the song too much, it will cause the system to operate slowly, and usually the application does not need to support so many files. If the customer has unconventional applications, please communicate with us in advance.

1).For example, select the first song played, serial transmission section: 7E FF 06 03 00 00 01 FF E6 EF
7E --- START command
FF --- Version Information
06 --- Data length (not including parity)
03 --- Representative No.
00 --- If need to acknowledge [0x01: need answering, 0x00: do not need to return the response]
00 --- Tracks high byte [DH]
01 --- Tracks low byte [DL], represented here is the first song played
FF --- Checksum high byte
E6 --- Checksum low byte
EF --- End Command

2).For selections, if choose the 100th song, first convert 100 to hexadecimal, the default is double-byte, it is 0x0064.
DH = 0x00; DL = 0x64

3).If you choose to play the 1000th, first convert 1000 to hexadecimal, the default is double-byte, it is 0x03E8
DH = 0x03; DL = 0xE8

4).And so on to the other operations, as in the embedded area in hexadecimal is the most convenient method of operating.

3.4.2 .Commands of Specify Volume

1). Our system power-on default volume is 30, if you want to set the volume, then directly send the corresponding commands.
2).For example, specify the volume to 15, serial port to send commands: 7E FF 06 06 00 00 0F FF D5 EF
3).DH = 0x00; DL = 0x0F, 15 is converted to hexadecimal 0x000F, can refer to the instructions of playing track section.

3.4.3 .Specify Device Play

1).The module default support four types of playback devices, the device must be on line, so it can specify playback. The software will automatically detect without user attention.
2).Refer the table as below to select the appropriate command to send
3).Module will automatically enter the Suspend state after the specified device, waiting for the user to specify a track playing. It will take about 200ms from specifying device to the module initialize file information. Please wait for 200ms and then send the specified track command.

Specify playback device –U-disk	7E FF 06 09 00 00 01 xx xx EF	xx xx: Verification
Specify playback device –TF Card	7E FF 06 09 00 00 02 xx xx EF	
Specify playback device -SLEEP	7E FF 06 09 00 00 05 xx xx EF	

3.4.4. Specify File to Play

Specify folder 01 of 001.mp3	7E FF 06 0F 00 01 01 xx xx EF
Specify folder 11 of 100.mp3	7E FF 06 0F 00 0B 64 xx xx EF
Specify folder 99 of 255.mp3	7E FF 06 0F 00 63 FF xx xx EF

1).Specify the folder playback is developed extensions, default folders are named as "01", "11" in this way because our module does not support Chinese characters identify the name of the folder name, in order to stabilize the system switching speeds and songs under each folder default maximum support up to 255 songs, up to 99 folders classification, if customers have special requirements, they need to classify according to the English name, we also can be achieved, but name only is "GUSHI", "ERGE" and other English name.

2).For example, specify "01" folder 100.MP3 file, serial port to send commands : 7E FF 06 0F 00 01 64 xx xx EF
DH: represents the name of the folder, the default support for 99 documents become 01 - 99 named
DL: on behalf of the tracks, the default maximum of 255 songs that 0x01 ~ 0xFF
Please refer to the above set rules for setting tracks

3).to the standard of the module, you must specify both the folder and file name, to lock a file. Individually specified folder or specify the file name alone is also possible, but the document management will be worse.

4).The following diagram illustrates both the folders and file names are specified.

01	folder name reference	2014/4/9 15:03	文件夹
11		2014/4/9 15:00	文件夹
31		2014/4/9 15:00	文件夹
99		2014/4/9 15:00	文件夹

Figure 3.1folder name

001.mp3	file name reference	2014/4/9 15:02	MP3 音频
002.mp3		2014/4/9 15:03	MP3 音频
255.mp3		2014/4/9 15:03	MP3 音频

Figure 3.2 file name

3.5. Key Ports

We use the AD module keys, instead of the traditional method of matrix keyboard connection, it is to take advantage of increasingly powerful MCU AD functionality, Our module default configuration 2 AD port, 20 key resistance distribution, if used in strong electromagnetic interference or strong inductive, capacitive load of the occasion, please refer to our "Notes."

1).Refer diagram

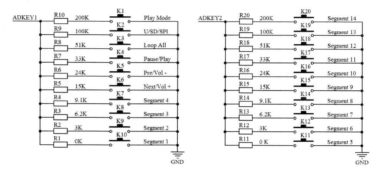

Figure 3.3 ad key refer

2)、20 function keys allocation table

Key	Short Push	Long Push	Description
K1	Play Mode		Switch to interrupt / non interrupted
K2	Playback device switches		U/TF/SPI/Sleep
K3	Operating Mode		All cycle
K4	Play/Pause		
K5	Previous	Vol+	
K6	Next	Vol-	
K7	4	Repeat play tracking 4	Long push always to repeat play
K8	3	Repeat play tracking 3	Long push always to repeat play

K9	2	Repeat play tracking 2	Long push always to repeat play
K10	1	Repeat play tracking 1	Long push always to repeat play
K11	5	Repeat play tracking 5	Long push always to repeat play
K12	6	Repeat play tracking 6	Long push always to repeat play
K13	7	Repeat play tracking 7	Long push always to repeat play
K14	8	Repeat play tracking 8	Long push always to repeat play
K15	9	Repeat play tracking 9	Long push always to repeat play
K16	10	Repeat play tracking 10	Long push always to repeat play
K17	11	Repeat play tracking 11	Long push always to repeat play
K18	12	Repeat play tracking 12	Long push always to repeat play
K19	13	Repeat play tracking 13	Long push always to repeat play
K20	14	Repeat play tracking 14	Long push always to repeat play

4、Application Circuit

4.1 Serial Communication Connect

Module's serial port is 3.3V TTL level, so the default interface level is 3.3V. If the MCU system is 5V. It is recommended connect a 1K resistor in series.

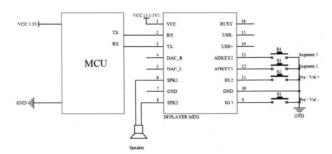

Figure 4.1 Serial Connect (3.3V)

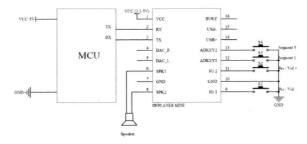

Figure 4.2 Serial Connect (5v)

4.2. Other Refer Diagram

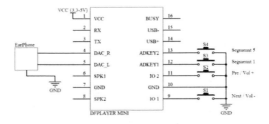

Figure 4.3 headset connect module

Between the headset and the module can string a 100R resistor, make a limiting

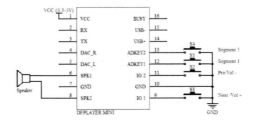

Figure 4.4 speaker connect module

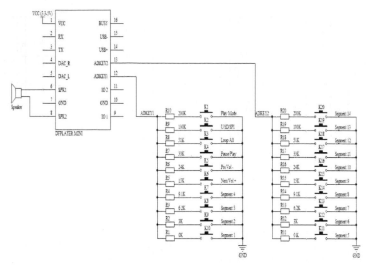

Figure 4.5 Ad key connect refer

5、MP3-TF-16P Size (unit: mil)

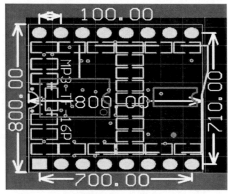

Figure 5.1 pcb size

6、Note*

I/O Input Specification

Item	Description	Min	Type	Max	Unit	Test Condition
VIL	Low-Level Input Voltage	-0.3	-	0.3*VDD	V	VDD=3.3V
VIH	High-Level Input Voltage	0.7VDD	-	VDD+0.3	V	VDD=3.3V

I/O Output Specification

Item	Description	Min	Type	Max	Unit	Test Condition
VOL	Low-Level Output Voltage	-	-	0.33	V	VDD=3.3V
VOH	High-Level Output Voltage	2.7	-		V	VDD=3.3V

1. The module's external interfaces are 3.3V TTL level, so please note the level conversion during the hardware circuit design, also in strong interference environment, electromagnetic compatibility note some protective measures, GPIO using opt coupler isolation, increasing TVS etc.

2, ADKEY key values are in accordance with the general use of the environment, if the strong inductive or capacitive load environment, please note that the module power supply is recommended to use a separate isolated power supply, another matched beads and inductors for power filtering, we must ensure that the input power as much as possible the stability and clean. If you really can not be guaranteed, please contact us to reduce the number of keys to redefine wider voltage distribution.

3. For general Serial communication, please pay attention to level conversion. If strong interference environment, or long distance RS485 applications, then please note that signal isolation, in strict accordance with industry standard design communication circuits.

資　　　　料　　　　來　　　　源　　　　　：

http://www.dfrobot.com/index.php?route=product/product&product_id=1121#.V6VWMPl97I

U

參考文獻

曹永忠. (2016). 物聯網系列：讓電腦發出音效（基本原理篇）. *智慧家庭*. Retrieved from https://vmaker.tw/archives/12788

曹永忠, 許智誠, & 蔡英德. (2014a). *Arduino 互动跳舞兔设计: Using Arduino to Develop a Dancing Rabbit with An Android Apps*. 台湾、彰化: 渥瑪數位有限公司.

曹永忠, 許智誠, & 蔡英德. (2014b). *Arduino 逆渗透滤水器控制器开发: Using Arduino to Develop a Controller of Reverse-Osmosis-based Water Purifiers*. 台湾、彰化: 渥瑪數位有限公司.

曹永忠, 許智誠, & 蔡英德. (2014a). *Arduino EM-RFID 门禁管制机设计:Using Arduino to Develop an Entry Access Control Device with EM-RFID Tags*. 台湾、彰化: 渥瑪數位有限公司.

曹永忠, 許智誠, & 蔡英德. (2014b). *Arduino EM-RFID 門禁管制機設計:The Design of an Entry Access Control Device based on EM-RFID Card* (初版 ed.). 台灣、彰化: 渥瑪數位有限公司.

曹永忠, 許智誠, & 蔡英德. (2014c). *Arduino RFID 门禁管制机设计: Using Arduino to Develop an Entry Access Control Device with RFID Tags*. 台湾、彰化: 渥瑪數位有限公司.

曹永忠, 許智誠, & 蔡英德. (2014d). *Arduino RFID 門禁管制機設計: The Design of an Entry Access Control Device based on RFID Technology* (初版 ed.). 台灣、彰化: 渥瑪數位有限公司.

曹永忠, 許智誠, & 蔡英德. (2014e). *Arduino 互動跳舞兔設計: The Interaction Design of a Dancing Rabbit by Arduino Technology* (初版 ed.). 台灣、彰化: 渥瑪數位有限公司.

曹永忠, 許智誠, & 蔡英德. (2014f). *Arduino 逆滲透濾水器控制器開發:The Development of a Controller for Reverse-Osmosis-based Water Purifiers Using Arduino* (初版 ed.). 台灣、彰化: 渥瑪數位有限公司.

曹永忠, 許智誠, & 蔡英德. (2015a). *Arduino 程式教學(入門篇):Arduino Programming (Basic Skills & Tricks)* (初版 ed.). 台湾、彰化: 渥瑪数位有限公司.

曹永忠, 許智誠, & 蔡英德. (2015b). *Arduino 程式教學(常用模組篇):Arduino Programming (37 Sensor Modules)* (初版 ed.). 台灣、彰化: 渥瑪數位有限公司.

曹永忠, 許智誠, & 蔡英德. (2015c). *Arduino 编程教学(常用模块篇):Arduino Programming (37 Sensor Modules)* (初版 ed.). 台灣、彰化: 渥瑪

数位有限公司.

曹永忠, 許智誠, & 蔡英德. (2015d). *Arduino 編程教学(入门篇):Arduino Programming (Basic Skills & Tricks)* (初版 ed.). 台湾、彰化: 渥玛数位有限公司.

曹永忠, 許智誠, & 蔡英德. (2016a). *Arduino 程式教學(基本語法篇):Arduino Programming (Language & Syntax)* (初版 ed.). 台湾、彰化: 渥瑪数位有限公司.

曹永忠, 許智誠, & 蔡英德. (2016b). *Arduino 程序教学(基本语法篇) :Arduino Programming (Language & Syntax)* (初版 ed.). 台湾、彰化: 渥瑪数位有限公司.

Arduino 程式教學（語音模組篇）
Arduino Programming (Voice Modules)

作　　者：曹永忠、許智誠、蔡英德

發 行 人：黃振庭

出 版 者：崧燁文化事業有限公司

發 行 者：崧燁文化事業有限公司

E-mail：sonbookservice@gmail.com

粉 絲 頁：https://www.facebook.com/
　　　　　sonbookss/

網　　址：https://sonbook.net/

地　　址：台北市中正區重慶南路一段六十一號八
　　　　　樓 815 室

Rm. 815, 8F., No.61, Sec. 1, Chongqing S. Rd.,
Zhongzheng Dist., Taipei City 100, Taiwan

電　　話：(02) 2370-3310

傳　　真：(02) 2388-1990

印　　刷：京峯彩色印刷有限公司（京峰數位）

律師顧問：廣華律師事務所 張珮琦律師

國家圖書館出版品預行編目資料

Arduino 程式教學 . 語音模組篇
= Arduino programming(voice
modules) / 曹永忠，許智誠，蔡英
德著 . -- 第一版 . -- 臺北市：崧燁
文化事業有限公司 , 2022.03
　　面；　　公分
POD 版
ISBN 978-626-332-079-6(平裝)
1.CST: 微電腦 2.CST: 電腦程式語
言
471.516　111001397

官網

臉書

定　　價：380 元

發行日期：2022 年 03 月第一版

◎本書以 POD 印製